PERGAMON INTERNATIONAL LIBRARY
of Science, Technology, Engineering and Social Studies
The 1000-volume original paperback library in aid of education, industrial training and the enjoyment of leisure
Publisher: Robert Maxwell, M.C.

Physics and Astrophysics

A Selection of Key Problems

Related Pergamon Titles of Interest

Books

DEMIANSKI
Relativistic Astrophysics

GINZBURG
The Propagation of Electromagnetic Waves in Plasmas, 2nd edition
Theoretical Physics and Astrophysics
Waynflete Lectures on Physics (Selected Topics in Contemporary Physics and Astrophysics)

JONES
The Solar System

PARRY
Essays in Theoretical Physics (In honour of Dirk ter Haar)

SINAI
Theory of Phase Transitions: Rigorous Results

Journals

Progress in Particle and Nuclear Physics
Progress in Quantum Electronics
Vistas in Astronomy

Full details of all Pergamon publications available on request from your nearest Pergamon office

Physics and Astrophysics

A Selection of Key Problems

by

V. L. GINZBURG

P. N. Lebedev Physical Institute of the Academy of Sciences of the USSR, Moscow, USSR

Translator

O. GLEBOV

Translation Editor

Gail ter Haar

PERGAMON PRESS

OXFORD · NEW YORK · TORONTO · SYDNEY · PARIS · FRANKFURT

U.K.	Pergamon Press Ltd., Headington Hill Hall, Oxford OX3 0BW, England
U.S.A.	Pergamon Press Inc. Maxwell House, Fairview Park Elmsford, New York 10523, U.S.A
CANADA	Pergamon Press Canada Ltd., Suite 104, 150 Consumers Road, Willowdale, Ontario, M2J 1P9 Canada
AUSTRALIA	Pergamon Press (Aust.) Pty. Ltd., P.O. Box 544, Potts Point, N.S.W. 2011, Australia
FRANCE	Pergamon Press SARL, 24 rue des Ecoles 75240 Paris, Cedex 05, France
FEDERAL REPUBLIC OF GERMANY	Pergamon Press GmbH, Hammerweg 6, D–6242 Kronberg-Taunus, Federal Republic of Germany

First English edition 1985

Translated from *O Fizike i Astrofizike: Kakie Problemy Predstavlyayutsya Seichas Osobenno Interesnymi*
published by Izdatel'stvo "Nauka"

Library of Congress Cataloging in Publication Data

Ginzburg, V. L. (Vitaliĭ Lazarevich), 1916–
Physics and astrophysics.
(Pergamon international library of science, technology, engineering, and social studies)
Updated translation of: O fizike i astrofizike. 3rd ed.
Bibliography: p.
1. Physics. 2. Astrophysics. I. Title. II. Series.
QC28. G5513 1984 530 83–23652

British Library Cataloguing in Publication Data

Ginzburg, V. L.
Physics and astrophysics
1. Physics 2. Astrophysics
I. Title II. O fizike i astrofizike. English
530 QC21.2
ISBN 0–08–026498–0 (Hardcover)
ISBN 0–08–026499–9 (Flexicover)

Printed in Hungary by Franklin Printing House, Budapest

Preface to the English Translation

Enormous numbers of books on physics and astronomy—textbooks, monographs, various collections of papers, and popular books—are published all over the world. But, to my astonishment, in recent years I have not seen a book similar to this present one. However, many readers would undoubtedly be interested in a general, albeit brief, review of many key problems of physics and astrophysics. Moreover, it would be desirable to have a choice of such books, since no single author can lay claim to his book being entirely objective and free of faults, to say nothing of a sufficiently detailed knowledge of the gigantic range of material to be covered.

At the same time, I must admit that the lack of other books of this type helps to reduce somewhat my anxiety about the publication of this translation. Unfortunately, there are still reasons for such anxiety and for feelings of dissatisfaction. Indeed, new problems arise continually and almost every week brings new data and new results. Therefore, the effective lifetime of each edition of this book is not long. Preparation of a revised and updated edition does not fully solve all the problems owing to the inertia of the previous editions, so to speak, which is difficult to overcome. In any case, if I wrote the book anew it would be different. I am particularly dissatisfied with the microphysical part of the book. It may be that in times to come the present period will be recognized as being no less significant than the years of the development of quantum mechanics. The gauge fields, the spontaneous breaking of symmetry, the quark model, the unified theories of different interactions—all these are now in the limelight and constitute an integral part of modern physics—but have not been adequately treated in this book. However, I hope that the book is, on the whole, modern and forward looking. Naturally, I have added new material while preparing it for translation. Most additions are placed at the ends of sections and are denoted by a bold asterisk. Besides being convenient, this method helps to identify and emphasize the most recent results.

In conclusion, I would like once more to stress my belief that books of this type are interesting and useful to many readers. So let my colleagues, particularly those who disagree with me on many issues, write other, better books.

January 1981

Unfortunately, the publication of the translation has been delayed. I have therefore added a few notes reflecting some results which, in my opinion, are important and which have been obtained in the last two and a half years. I also note that, if I were to be writing this book now, I would also include amongst the key problems one concerning solitons, strange attractors and "chaotic" solutions which have been obtained in the analysis of a number of non-linear equations which describe many physical systems.

June 1983 V. L. GINZBURG

Preface to the Third Russian Edition

Five years have passed since the previous edition of this book. Only under exceptional circumstances do profound changes in science occur within such a short period. Such was the period between 1925 and 1930, for instance, when quantum mechanics was created, and largely developed. The last five years have not been exceptional for physics and astronomy. However, much has been done in this period, and of course this book should reflect the current state of the problems with which it deals. In general, irrespective of the success or usefulness of the present book, which the author has no right to judge, there is clearly a marked interest in literature of this type, as evidenced for example by translations of this book into English, French, German, Polish, Slovene and Bulgarian.

Since this is a new edition of the book, but with a similar title, I have been restricted, to a certain extent, by the previous editions. Therefore, I have only made additions and changes which concern the essence of the physical and astronomical problems discussed in the book.

In order to do this, I have reviewed a large number of new papers in a wide variety of fields, and in the process have realized how difficult it was to cover, even superficially, a considerable part of modern physics and astrophysics. In this connection, I should like to stress once more that I have never regarded this book as something outside the scope of popular presentation. To evaluate it as if it were a programme document or a philosophical treatise would mean losing the sense of proportion. Apparently I lost it myself in heatedly answering criticisms which I considered unjustified. I still believe that, with the above reservations, we can identify "the most interesting and important" problems, and we can and must discuss the relative significance of various research fields. I also believe that the author of such a book need not bear in mind the possible views of higher authorities or the special interests of some of his colleagues. On the other hand, this controversy is somewhat "outdated" now and, so as not to annoy my critics, I could smooth over some points and, for instance, write about "certain important and interesting problems" instead of "the especially important and interesting" ones. Thus, if I wrote the book anew it would look different. However, I have not made changes in this direction, and have preserved all the general discussions and remarks, which are sometimes rather controversial. As I said, this is not the first edition; the author has nothing to lose, and the heated or even controversial character of the discussion can only make the reading more interesting.

In conclusion, I wish to thank all my colleagues whose advice was useful in the preparation of this edition.

June 1979 V. L. GINZBURG

Preface to the Second Russian Edition

Physics has grown and diversified immensely in recent decades; this is demonstrated by the emergence of such new sciences as astrophysics, biophysics, geophysics, chemical physics, physics of crystals, physics of metals, etc. This differentiation, however, has not deprived (or perhaps it would be more correct to say has not yet deprived) physics of a certain unity. I mean here the unity of the fundamentals, the generality of many principles and methods, as well as the bonds between various branches and fields of research. At the same time, differentiation and specialization are increasingly hindering visualization of the structure of physics as a whole, leading undoubtedly to some disunity. This disunity seems to be, to a certain extent, inescapable, but the desire to compensate somehow for its consequences is quite justifiable.

This is particularly significant for young physicists, and, primarily, for students. It is a fact that even the best graduates of the physical (and related) departments of our universities lack an overall view of the present situation in physics as a whole, having specialized in a more or less narrow field of it. Of course, one cannot get a "bird's eye view", or at least versatile knowledge, overnight, and a university training can hardly achieve these goals.

But sometimes the lack of consistency and even lapses of knowledge are truly astonishing. For instance, a person may know fine modern methods of quantum statistics or quantum field theory but have no understanding of the mechanism of superconductivity, or the nature of ferroelectricity; he may not even have heard of excitons or metallic hydrogen; may be unaware of the concepts of neutron stars, "black holes", gravitational waves, cosmic rays and gamma radiation, neutrino astronomy and so on. I believe that the reasons for this are not human limitations or the lack of time. It would perhaps take less time and effort for a student to get a basic physical "picture without formulae" of all the above and similar subjects (or, at least, with the use of only the simplest formulae and quantitative concepts) than to prepare for a major examination.

The difficulty lies elsewhere—the student does not know with what to get acquainted, and how to do it. It is not enough for certain subjects to be mentioned in some of the numerous university courses or text-books.

Moreover, the very problems that get most attention at physical conferences or in the journals are too novel to find their way into text-books or university curricula.

It is hardly necessary to dwell on the subject and the conclusions seem to be self-evident. If we limit ourselves to discussing our good intentions or calling for the improvement or updating of university courses, frequently our goal will never be reached. The most reasonable solution seems to be to deliver regular additional lectures for students, according to a special schedule (8–10 lectures a year), that are not included in any of the established courses. Each lecture should be delivered by an expert in the respective field. These extracurricular lectures should each be a review, simple, but up to date, of a certain research field or problem.

The Chair of Problems in Physics and Astrophysics at the Moscow Institute of Physics and Technology had scheduled a series of such lectures. But these lectures had to be preceded by some sort of a general introduction, a "bird's eye view", an unavoidably fragmentary and cursory review of many problems, an attempt to present the current problems in physics as a whole. This task seems to be a difficult and, in a sense, not a gratifying one, as its fulfilment can hardly be successful enough. Anyway, usually nobody gives such lectures. But since I considered such a lecture to be a prerequisite for the success of the above lecture series as a whole, I got down to work on it. The lecture was later delivered on a number of occasions to various audiences. The way it was received demonstrated unambiguously that such lectures are, to say the least, necessary and attractive—and not only for students. The lecture eventually developed into a paper entitled "What problems of physics and astrophysics seem now to be especially important and interesting" that was published in the "Physics of Our Days" section of the journal *Uspekhi Fizicheskikh Nauk*, **103,** 87 (1971) and then translated into a number of languages and published as a book by Znanie Publishers (1971).* The present small book is an extended and updated version of that paper; it contains a few new sections and some other alterations. Such alterations were rendered necessary, in particular, by the accumulation of new data. It is hardly necessary to discuss the contents of the book in more detail here; one can get acquainted with it by looking through the list of Contents and Introduction.

There are reasons for such a lengthy preface to so small a book. These are that its contents, character and style seem to be somewhat unconventional or at least not self-explanatory. I have addressed my book to budding

* After this English edition had been prepared I published a paper with the same title but subtitled "Ten years after" (*Sov. Phys. Usp.*, **134,** 469, 1981). This paper presents my views on the development of physics and astrophysics in the last decade.

physicists and astronomers; I have stressed that the selection of the "most important and interesting problems" is tentative and subjective in character; I have also noted that any evaluations under such circumstances inevitably become controversial but, at the same time, that I am far from having any bias or pretensions to preach, or to impose my opinions on the readers. Fortunately, as far as I have gathered, the paper has been accepted in just this way by the majority of readers, especially by those to whom my message was addressed. But opposite opinions have also been voiced. Some did not like the very idea of the paper. Others considered it to be intolerably biased, especially against microphysics (I have even been granted the title of the "enemy of nuclear physics"). Still others charged me with a lack of modesty and suchlike sins that they inferred from my attempts to judge what is important and what is not, as well as from too frequent appearances of my name in the Bibliography which plays a purely auxiliary role in the paper. It would be inappropriate to answer all these allegations and reproaches here, all the more so since they, unfortunately, have not been published anywhere. But they are worth mentioning in order to caution readers against the possible dangers to which they might be exposed, and thus to stimulate a critical approach to their reading. I myself have tried my best in this respect when preparing the present edition. But to pay attention to criticism does not mean to "fear the clamour of the Boeotians" and drop a cause which seems to be worthwhile.

As is clear from the above, I am interested in the opinions of as many readers as possible. I would be grateful for letters of criticism, suggestions and general remarks about the book. I am thankful to those whose advice has been used when this edition was prepared.

1973 V. L. Ginzburg

Contents

Introduction

Physics and astrophysics are nowadays concerned with an enormous number and variety of problems. Attempts at solving these problems are usually worthwhile, as they allow physicists if not to uncover the secrets of nature, then at least to gain new knowledge. None of these problems may rightly be thought of as devoid of interest or importance. However, there does exist a hierarchy of problems which is reflected in all scientific (and sometimes not only scientific) activities. "Especially important" physical problems are often identified according to the potential effect they may have on either technology or the economy, a special fascination of the problem, or its fundamental character. Sometimes the choice is due to fashion, or to other obscure or hazardous factors. We shall, of course, avoid discussion of the latter type.

This is not the first time that a list of "most important problems" has been compiled and commented upon. Conferences are often convened or special commissions set up to do this. These may produce bulky reports. I do not intend to generalize, but must say that I have never seen anyone reading such reports on "most important problems" with any great interest. It seems that specialists do not really need them, and they do not attract a wider reading public. Of course, such documents may prove to be necessary for the planning and financing of the development of science.

And yet, physicists and astronomers—especially young ones—tend to ask a simple question: what is "hot" in physics and astrophysics? Or, in other words, what seems to be most important and interesting in physics and astrophysics at present? Assuming that a sufficiently large number of readers are interested in this question, I have attempted to answer it in this small book. The book is not a product of a commission's deliberations and not even a result of special investigations. It is, rather, the author's personal view. This, at least, makes it possible to avoid the dry and bare style of the more official documents.

The problems that seem to me now to be especially important and interesting are listed below. At the same time, I do not attempt to justify my selection criteria. Everyone has a right to their own views, and should not feel obliged to co-ordinate them with those of anyone else unless he or she declares his or her views to be authorized or superior to others. I attempt nothing of the kind and, of course, make no organizational suggestions; in order to stress this personal approach I have not even tried to avoid using personal pronouns as is customary in scientific papers.

It would be interesting and perhaps instructive to compare the lists of the "most important problems of physics and astrophysics" compiled by a number of people. Unfortunately no such poll of scientists' opinions has ever been conducted, as far as I know. Therefore I can only suggest that the majority of such lists would have many elements in common, provided that the following difficult requirement is met: that a consensus is reached in defining the "physical problem" concept as distinct from, say, fields, trends, or objects of physical studies. By a problem I mean a question, the answer to which is substantially unclear in character and content. We should deal not with technological developments, measurement projects, etc., but with the possibility of creating some new substance with unusual properties, say a high-temperature superconductor, establishing the limits of applicability of a theory—for instance, the general theory of relativity—or throwing light on something really unknown; for example, the mechanism for the breaking of combined parity in the decay of K-mesons. This is precisely the reason why, in this book, I practically ignore quantum electronics (including the majority of laser applications) as well as many problems in the physics of semiconductors (including miniaturization of circuits and devices), non-linear optics and holography and some other interesting trends in the development of modern optics, the problems of computer technology (including the development of novel types of computers) and many other problems.

These issues are, undoubtedly, very important and have many technological and physical implications. However, they do not involve any essential "physical problem" or any basic "uncertainty" concerning the underlying physics. There was, for example, such an uncertainty prior to the development of the first laser, even though the principles which were used later for laser design had been known. Increasing the power, or changing other parameters, of a laser or any other device may be a necessary, difficult and commendable task but is, of course, qualitatively different from developing a device or a machine on the basis of new principles. At the same time, this is a fairly typical example for illustrating the arbitrary nature of the boundary between the basic research problems and the technological problems in physics. For instance, increasing the laser power by many orders of magnitude (although a currently important problem) cannot be classified as a purely technological or some kind of "non-basic" task. The same may be said about the development of X-ray "lasers" and grasers, the laser analogues of X-rays and gamma rays. X-ray lasers and grasers have not yet been developed and it is still unclear as to how this will be done, or whether such development is feasible: thus, they present a typical "important and interesting problem" in terms of our selection rules. The same is true for almost any field—a significant breakthrough almost always constitutes a problem. But not all such problems are ripe enough for solving; not all prizes seem tempting and there exists in fact a hierarchy of problems.

At the same time, we cannot, of course, deal only with individual problems, however important, and ignore the wide variety of other tasks and problems which failed to make the grade of "important and interesting problems". Moreover, these problems can prove to be both very difficult and very interesting, at least, for those who work on them. I can illustrate this argument with problems from the theory of radiation from sources travelling through a medium (Cherenkov radiation, transition radiation, transition scattering, etc.). I am greatly attached to and fascinated by this field as I have been working in it throughout my research career[(1)]. But one cannot help seeing that such problems in electrodynamics involve no real mysteries and in this respect they differ substantially from, say, the problem of high-temperature superconductivity or the problem of quarks and their confinement in the bound state. It is natural, therefore, that this book deals neither with transition radiation nor with a number of other problems in which I am, or have been, interested. Thus, though this selection of the "important and interesting" problems still is, in a sense, arbitrary and subjective, it is by no means based on the principle that the important and interesting problems are primarily those on which the author is working (I do not think that this remark is superfluous since one fairly often meets people who use precisely this selection principle).

It has been suggested above that a "poll of scientific opinion", if attempted, would show a large measure of agreement on the selection of current "especially important and interesting problems". However, significant disagreements would also be inevitable, especially as far as priorities in allocation of resources and concentration of efforts are concerned. This is clear, for instance, from the literature[(2–5)].

The question of resources and priorities is, however, linked to a variety of factors lying outside the scope of purely scientific problems. For instance, the construction of mammoth accelerators is undoubtedly of great scientific interest, but what is argued is whether the expenditure involved produces results that may justify the necessary curtailment of research in other areas. We shall ignore this aspect of the discussion and concern ourselves only with scientific issues. However, even with this "simplification" and restriction, opinions may diverge sharply. For instance, the most important problems of solid state physics are listed here: high-temperature superconductivity, the creation of metallic hydrogen and some other materials with unusual properties, metallic exciton liquid in semiconductors, surface effects and the theory of critical phenomena (in particular, the theory of second-order phase transitions). An article[5] entitled "The most basic unsolved problem in solid state physics" states that this problem is to explain the empirical formula for the heat of formation of some crystals from other substances. With some effort, I found an interest in this problem but I failed completely to understand why the problem was thought to be the "most basic" one—and, what is more, I greatly doubt it. What is the conclusion? There seems to be only one possibility: no authoritative lis tof the most important problems can be suggested; and there is no need for one. But it is both necessary and useful to evaluate what is important, and what is not, to argue about it, and to be bold in putting forward suggestions and defending them (but not to impose one's own views). This is precisely the spirit in which this book is written.

Thus, the subjective and controversial character of this book is quite apparent and readers have been warned (although, of course, such warnings are rarely heeded). It is only left to note that the division of the book into three parts—Macrophysics, Microphysics, and Astrophysics—is quite arbitrary. For instance, we discuss superheavy nuclei under the heading of macrophysics, though they may be said to constitute a microphysical problem. Furthermore, the problems of the general theory of relativity are treated in the astrophysical part, rather than among macrophysical problems. The only reason for this is that this theory is used mainly in astronomy (to say nothing of the fact that the difference between astrophysics and, say, macrophysics is essentially of quite another character than the difference between macrophysics and microphysics).

Finally, it should be noted that the book practically ignores biophysics, let alone other less important research areas related to physics and astrophysics. However, it is precisely the co-operation between physics and biology and the application of physical methods and concepts that have proved to be especially fruitful and significant in the development of biology, medicine, agricultural sciences, and so on. It would be a gross error for physicists to avoid "biologically biased" problems on the grounds of their not being "physical" (this has been convincingly argued elsewhere[2]). Moreover, it is conceivable that this co-operation with biology, and attempts to solve biological problems, will stimulate the development of physics proper, just as physics was, and still is, a source of inspiration and new ideas for many mathematicians. Thus, even though this book does not pay due attention to the links between physics and the biological sciences, this does not reflect an underestimation on my part of their importance, but rather my inadequate knowledge of biophysics and biology in general and, also, the necessarily limited scope of this book.

I.

Macrophysics

1. Controlled thermonuclear fusion

The solution of the problem of controlled thermonuclear fusion implies the use of the nuclear fusion reactions for power production. The following basic reactions are involved:

$$\left.\begin{aligned} d+d &\rightarrow {}^{3}He+n+3.27\ MeV\\ d+d &\rightarrow t+p+4.0\ MeV\\ d+t &\rightarrow {}^{4}He+n+1.76\ MeV \end{aligned}\right\} \qquad (1)$$

(here d and t are the nuclei of deuterium and tritium, p is the proton, n is the neutron).

Another important reaction is

$$^{6}Li+n \rightarrow t+{}^{4}He+4.6\ MeV$$

since it gives rise to tritium, which does not occur naturally. Some other reactions may also prove to be useful; for example, the reaction

$$d+{}^{3}He \rightarrow {}^{4}He+p+18.34\ MeV$$

It can scarcely be questioned that nuclear fusion energy will be used in some way or other: one has only to mention the "obvious" possibility of useful underground explosions. On the other hand, controlled thermonuclear fusion has been attracting great attention for 30 years although a thermonuclear energy "yield" exceeding the thermal plasma energy has still not been obtained. However, installations are now being built, and which are to be tested during 1982–1985, as prototypes for the real thermonuclear reactor. According to some predictions, a commercial reactor will have been built by the end of this century or the beginning of the next.

In order to make the thermonuclear energy yield higher than the energy consumed for plasma heating, the condition $n\tau > A$ must be satisfied, where n is the electron concentration* in the plasma at a temperature

*Of course, plasma is fully ionized at the high temperatures needed for the reactor operation ($T \gtrsim 10^8$ K) and the concentration of electrons is approximately equal to the concentration of deuterium and tritium ions. The equality is approximate since plasma always contains impurities—carbon, oxygen, etc. (see, for example, reviews[6]).

$T \sim 10^8$ K and τ is the characteristic time of plasma confinement (for instance, it may be the time during which the energy lost by plasma is of the order of its thermal energy). The constant A characterizes the nuclear fuel (and the concentration of the impurity atoms). For pure deuterium $A \sim 10^{16}$ cm^{-3} s and for a mixture of 50% deuterium and 50% tritium $A \sim 2\times10^{14}$ cm^{-3} s (A can be decreased by a factor of almost 10 by using the neutrons produced during thermonuclear reaction for fission of uranium). Thus, in order to make a reactor function (the energy it produces must be greater than the energy needed to establish and maintain high plasma temperatures) in the "pure" reactor, that is one which does not contain fissionable materials (uranium, etc.), we need to satisfy the following condition

$$n\tau > 10^{14}\ \text{cm}^{-3}\ \text{s} \tag{2}$$

The physical meaning of this condition (2), known as the Lawson criterion, is clear enough—the longer the time of reaction, the lower the fusion reaction rate—it is proportional to n^2.

Magnetic confinement of plasma might appear to be the simplest approach to the plasma reactor design. Among reactors of this type the most well known and popular are the toroidal magnetic traps—tokamaks. In 1979 a record value of $n\tau = 3\times10^{13}$ cm^{-3} s was obtained in the MIT tokamak. This reactor was relatively small, but had a strong magnetic field of up to 90 kOe. The plasma temperature was about 10^7 K and the plasma concentration at the centre was up to 10^{15} cm^{-3}. In the T-10 tokamak (built in the Kurchatov Atomic Energy Institute in Moscow) the energy lifetime is about 0.06 s and the ion temperature is about 1.2×10^7 K. Similar values have been obtained with the PLT tokamak, the largest in the USA. A plasma temperature of 6×10^7 K was obtained in this machine in 1978. Construction of the test reactor tokamak FTRT using a deuterium–tritium mixture was started in 1977 in the USA. Other machines to be launched soon are the European tokamak in Britain, the DT-60 tokamak in Japan, and the T-10M tokamak in the USSR. The plasma volume in such tokamaks is over 100 m^3. They will, probably, make it possible to reach the value of $n\tau \sim 10^{14}$ cm^{-3} s before 1986. The cost of the FTRT machine is much greater than 200 million dollars. The next step is the development of a power reactor with the circulation of tritium, which should produce thermal energy. The development costs at this stage will be very high and discussions are now under way to review the feasibility of an international project for the development of the power reactor tokamak.

The magnetic field of the fusion reactor should be produced by superconducting coils; otherwise a favourable energy balance cannot be expected. The recently launched tokamak T-7 has superconducting coils, and the future tokamak T-15M will also have them. But many of the physical and technological problems of reactor operation have not yet been solved. They include the problem of the durability of the first wall of the reactor which is irradiated by a high-intensity neutron flux. The problem of plasma heating has also not been solved. The fact is that ohmic heating by itself is not sufficient to obtain the required plasma temperature. Work is under way to test the techniques of plasma heating by beams of neutrals (deuterium atoms with energy 20–100 keV) or microwaves. Moreover, we have an inadequate knowledge of the behaviour of the impurity atoms in tokamaks and of the causes of the high electronic heat conductivity.

Significant advances have been made with open-ended magnetic traps using magnetic mirrors. The plasma parameters in these can be as high as $T \sim 10^8$ K and $n \sim 10^{14}$ cm^{-3}. The lifetime τ in such systems, however, is as low as 0.001 s, making for a low value of $n\tau \sim 10^{11}$ cm^{-3} s. The reason for this is that in the magnetic traps even one collision of one ion with another removes an ion from the system. Perhaps plasma confinement in the traps will be improved by modification of the magnetic mirrors at the ends.

The above difficulties, which can prove even greater in real systems, justify attempts to devise other approaches to the problem. Therefore, apart from tokamaks and magnetic traps, other systems and techniques such as stellators, high-frequency discharges in plasma, compression of shells thus creating magnetic fields of the order of a million oersteds, etc., are being tested and discussed.

In recent years, considerable attention has been focused on studies of the possibility of inertial confinement fusion. This method involves the use of micro-explosions accompanied by the liberation of energy up to 10^8 J. (For instance, a deuterium–tritium pellet, about a millimetre in size, produces energy of the order of 3×10^8 J in the case of complete fusion. This corresponds to the energy liberated by the explosion of about 50 kg of TNT.) The destructive effect of such an explosion is relatively small since the mass of exploded material is small, and hence the momentum is also small. Since the time of energy loss for the explosion is of the order of 10^{-8}–10^{-9} s, the heating power should be about 10^{14} W (see below). In principle, such a high heating power can be obtained either with a laser beam, with an electron beam, or with a beam of ions. Accordingly, the fusion systems discussed are known as the laser, electron, and ion beam systems. Of course, the mechanisms of absorption of electrons, ions and laser radiation by the target (the fusion fuel) are different, but if we ignore this difference we can readily see the similarity between the above methods. Indeed, whether we use laser radiation, electrons or ion beams to heat plasma, we have to irradiate as homogeneously as possible, solid spherical pellets of hydrogen (or, more exactly, deuterium or a deuterium–tritium mixture), at an initial concentration of the nuclei of $n \sim 5\times10^{22}$ cm^{-3}. This is the concentration of nuclei in solid hydrogen at atmospheric pressure. The nuclear fuel is sheathed with a number of shells known as pushers, rammers, rammer pushers and ablators. When the outer shell (the ablator) evaporates, it produces a pressure of up to 10^{12} atm, resulting in a compression of the nuclear fuel by a factor of 1000 or more. Of course, the structure of the shells and of the target pellet is chosen to provide the highest degree of compression of the fuel. The most important requirement is that the alpha particles produced in the fuel be retained in the target and maintain the reaction. It should be borne in mind here that the mean free path of the particles decreases proportionally with increasing fuel density, while the rate of decrease of the pellet radius is considerably lower (proportional to $n^{1/3}$).

The main problem for inertial confinement fusion systems lies in obtaining a large ratio, Q, between the liberated fusion energy and the energy of the light, electron or ion beam fed into the pellet. As estimated, Q may be as high as 60–70; then, in order to obtain a positive energy yield in the system, the laser efficiency should be as high as 10–20%. The efficiency of currently available lasers, which produce nanosecond pulses, is less than 1%.

Another important requirement is highly durable laser materials. It has been estimated that the laser glass must withstand 10^8 pulses before failing, but the lifetime of available materials is shorter by a factor of 10^4. Of course we can attempt to continue the reaction in the pellet, and thus not at the expense of further laser heating, by means of self-maintenance (that is, by further heating with alpha particles). Q values of a few hundreds can, apparently, be obtained in this way, and the laser efficiency required can thus be lower. But this approach also has a number of difficulties related to the development of instabilities in the shells, the generation of fast electrons, and so on. Nevertheless, scientists hope to carry out a demonstration experiment soon. (This demonstration involves a fusion reaction with $Q = 1$, so that the fusion energy yield is equal to the energy consumed in heating the fuel.) The large-scale laser fusion installation Shiva (Livermore, USA) started operation in 1977. In the Shiva installation 20 laser beams feed the target about 10 kJ of energy. The first experiment with the Shiva installation was performed in 1978.

Laser fusion installations are being built and designed in the USSR at the Lebedev Physical Institute in Moscow (Delfin, UMI-35), and in other countries. The fusion installations Angara-5 (Kurchatov Atomic Energy Institute, USSR) and EBFR (Sandia Laboratory, USA) will use electron beams. Work has been started on the design of fusion installations using ion beams. The expenditure for fusion projects in the USA was 500 million dollars in 1979.

Enormous difficulties still remain in the development of fusion reactors with magnetic plasma confinement, or inertial confinement reactors. Nevertheless, at present, in contrast to the fairly recent past, the general feeling is one of optimism, and it seems to be basically possible to develop some kind of fusion reactor. But what type or types of reactors it will be possible to build, when this will be done, and what difficulties remain to be overcome—the answers to all these questions are by no means clear. Moreover, the difficulties involved are so significant that they cannot be regarded as purely technological. Therefore, the development of fusion reactors should be classified as one of the most important physical problems. Also, there seems to be a clear need for competition between the various approaches to the problem of controlled thermonuclear fusion (by this I mean fair competition, not rivalry).

Incidentally, the following general principle is clearly exemplified by the problem of controlled thermonuclear fusion: practically no large-scale physical problem stands apart from all others, but will be closely related to a variety of different branches or fields of physics. Therefore, the especially strenuous efforts made in solving a given problem may bear fruit in a more general sense—they may stimulate new studies, give rise to novel methods and approaches, and so on. For instance, plasmas had attracted considerable scientific interest even before the early 1950s when the problem of controlled thermonuclear fusion emerged. But one can hardly overestimate the importance of the results of plasma physics obtained in this field for gas, solid state and space plasmas.

*Looking through the papers published during the last four Years, I could find no dramatic news of controlled thermonuclear fusion. Tokamaks are still favourite, but the interest in stellators has again increased (they differ from tokamaks in the additional coils that produce the azimuthal magnetic field). The work on open-ended magnetic traps continued in the hope of developing improved magnetic mirrors. It can hardly be predicted that the open-ended systems (which are the simplest and most convenient in some respects) will never compete with toroidal systems. Naturally, the work on the theoretical and practical aspects of inertial confinement systems is also continuing, with attention being focused on heating the fuel with laser or ion beams.

2. High-temperature superconductivity

The phenomenon of superconductivity was discovered in 1911, and for many years it remained not only unexplained (perhaps the most puzzling phenomenon in macrophysics) but also useless practically. This latter fact is largely because, up till now superconductivity has only been observed at low temperatures. For instance, the first superconductor discovered—mercury—has a critical temperature T_c of 4.15 K. One alloy of Nb, Al and Ge was found fairly recently to have one of the highest T_c values, about 21 K. In 1973 the compound Nb_3Ge was found to have $T_c = 23.2$ K (there is a better known superconducting compound, Nb_3Sn, with $T_c = 18.1$ K, which was discovered in 1954).

The use of superconductors becomes especially difficult around the critical temperature (of course, we mean below T_c since, by definition, a metal ceases to be superconducting at higher temperatures). Suffice it to say that in this temperature region the critical magnetic field, H_c, and the critical current, I_c (that is, the field and current that destroy superconductivity) are very small. When T tends to T_c the values H_c and I_c tend to zero. Thus superconductors can be used only when cooled by liquid helium (boiling point at atmospheric pressure $T_b = 4.2$ K) since liquid hydrogen (boiling point $T_b = 20.3$ K) freezes at 14 K and it is generally difficult and inconvenient to use solids for cooling.

As recently as 30 years ago the production of helium was low (it is not sufficient, even now) and liquefaction techniques were inadequate. Only a small number of low capacity helium liquefiers were operating in the world. The use of superconductors for the construction of superconducting magnets (which is the most important application so far) was limited to a no lesser extent by the low values of H_c and I_c of materials available at the time (for Hg the critical field is about 400 Oersted (Oe) even at temperatures tending to zero).

However, things changed radically at the turn of the 1960s. Liquid helium is now readily available. Where it is done properly, laboratories do not install liquefiers; instead, they order by phone the required amounts of liquid helium from specialized firms and helium is shipped in large Dewar vessels. The "magnetic and current barrier" has also been overcome; superconducting materials now available make it possible to build magnets with a critical field as high as hundreds of kilo-oersteds (the above-mentioned alloy of Nb, Al and Ge, which has a critical temperature of 21 K, has a critical magnetic field of about 400 kOe; the record observed value of H_c is about 600–700 kOe). It is true that materials currently used have critical magnetic fields and currents too low for a 300–400 kOe magnet to be built, but this seems to be a purely technological problem. In principle, there appears to be no fundamental factor preventing the construction of, say, a 300 kOe magnet operating at helium temperatures. (Superconductors with high H_c and I_c values have, basically, been the result of large-scale research and development work. No decisive role has been played here by theoretical studies, particularly in the effort to obtain higher critical currents, but other advances have been initiated by theoretical developments. Thus, roads to success may be essentially different, depending on circumstances.)

A fundamental, vague problem in superconductivity is the extremely attractive possibility of creating high-temperature superconductors; that is, metals that become superconducting at liquid nitrogen temperatures, or better, at room temperature. (The boiling point for nitrogen is $T_b = 77.4$ K.) I have discussed the current state of high-temperature superconductivity in detail elsewhere[7, 8]; therefore, only a few remarks are made here.

Superconductivity appears in metals when electrons in the vicinity of the Fermi surface are attracted to each other, thus producing pairs which undergo something like the Bose–Einstein condensation. The critical temperature T_c for the superconducting transition is proportional to the bonding energy of the electrons in a pair, and is determined, roughly speaking, by two factors—the force of attraction (bonding), which may be described by a factor g, and the width $k\theta$ of that energy range near the Fermi surface where there still exists attraction between electrons. We have

$$T_c \sim \theta \exp(-1/g) \tag{3}$$

The majority of known superconductors have $g \lesssim 1/3$–$1/4$ (formula (3) may be used directly only for $g \ll 1$). The temperature θ in (3) depends on the mechanism of attraction between the electrons. In the known superconductors this mechanism seems to be due to the interaction between electrons and lattice. Then we have $\theta \sim \theta_D$ where θ_D is the Debye temperature whose physical meaning is illustrated by the fact that $k\theta_D$ is the energy of the phonons with the shortest wavelength in the solid ($k = 1.38 \times 10^{-16}$ erg/K, and is the Boltzmann constant). This wavelength is $\lambda \approx a \approx 3 \times 10^{-8}$ cm (a is the lattice parameter). $k\theta_D \sim \hbar\omega_D$ (here $\omega_D \sim u/a \sim 10^{13}$–$10^{14}$, where $u \sim 10^5$–10^6 cm/s is the speed of sound). Thus we obtain $\theta_D \sim 10^2$–10^3 K.

When $\theta_D = 500$ K and $g = 1/3$ we have, according to (3), $T_c \sim \theta_D e^{-3} = 25$ K, and, in general, $T_c \lesssim 30$–40 K for the phonon mechanism (this can also be shown by considerably more detailed analysis[8]). Thus, on the one hand it still seems to be possible that T_c can be increased by conventional methods (by preparation of new alloys and by special treatment of them, not to mention compounds of the type of metallic hydrogen (see below)). On the other hand, it is clear that it would be difficult, if not impossible, to create a real high-temperature superconductor with $T_c \gtrsim 80$–300 K on the basis of the phonon mechanism (here again we do not take into consideration metallic hydrogen).

Hopes of obtaining high-temperature superconductivity are based primarily on the use of the exciton mechanism of attraction between electrons. It is well known in solids that, apart from lattice waves (in quantum language, phonons), electronic excitations known as excitons may appear. In molecular crystals, excitons are represented by an excited state of a molecule which jumps from one molecule to another and thus travels along the crystal. In semiconductors, the simplest example of an exciton is an electron and hole bonded by Coulomb forces, thus forming a quasi-atom similar to the positronium atom. The excitation energy (bonding) of such excitons (here we mean excitons of the electronic type, as other types of excitations are sometimes referred to as excitons) ranges typically from a few hundredths of an electron volt to a few electron volts. As with phonon exchange, the exchange of excitons can give rise to attraction between conduction electrons. But in such cases we obtain a temperature of $\theta \sim E_e/k \sim 10^3$–$10^5$ K from a formula of the type (3). Here E_e is the excitation energy; the energy $E_e \sim 1$ eV corresponds to a temperature of $\theta \sim 10^4$ K. Therefore, if the exchange of excitons could provide a sufficiently strong attraction between electrons ($g \gtrsim 1/4$–$1/5$) the critical temperature would be high. Several approaches have been suggested to make use of the exciton mechanism of superconductivity. One such approach is to use layered compounds or "sandwiches" of thin metal layers. For a long time (from 1964) I regarded this approach as the most promising one.

Very interesting superconducting layered compounds have been obtained[(8)] but their critical temperatures, as with that of sandwiches, are not high enough. One can hardly say that the scope of studies in this field is impressive, especially when one compares it with those of the fusion reactor development effort, or accelerator construction work. The reason apparently is that the theory cannot put forward simple specific suggestions as to how to find high-temperature superconductors and, most importantly, cannot guarantee success. On the other hand, maybe we do not need to perform ultracomplicated synthesis of new compounds to obtain high-temperature superconductors; it may be that success could be achieved with comparatively modest (although advanced) methods. Therefore, I would not be too surprised to read about the creation of a high-temperature superconductor in a current issue of a physical journal (though that would probably be a sensation, and we would hear about it in the popular press). It is equally probable that the creation of high-temperature superconductors is very difficult, or even impossible, in principle. As usual in such cases, evaluation of the possibilities of success vary from the hopeful[7, 8] to the quite pessimistic[9].

Between 1977 and 1979 the following advances were made in this field. It has been shown theoretically[10] (see also[8]) that the general objections[9b] to the possibility of obtaining high critical temperatures are incorrect. Thus, we can now say that there are no fundamental objections to the estimate of $T_c \lesssim 300$ K, that is, in principle, to the possibility of creating high-temperature superconductors. At the same time, it becomes increasingly clear that even if we can reach this goal this can only be done under rather special conditions.

On the experimental side I should make a special mention of the discovery of the metallic conductivity (and superconductivity with $T_c \approx 0.3$ K) of the polymeric sulphur nitride $(SN)_x$ which, of course, contains no metal atoms. This finding proves that a non-zero conductivity at $T \to 0$ (that is, by definition, metallic conductivity) can be found in a considerably wider range of materials than was earlier assumed. The search for new materials possessing metallic conductivity and superconductivity may yield especially interesting results from compounds containing light atoms (in particular, organic compounds) because there are reasons to expect higher critical temperatures for them[8].

Apart from unsubstantiated reports of observations of superconductivity at fairly high temperatures which we, naturally, shall not discuss here, one of the sensations of 1978 was the report[11, 12] of the discovery of very high diamagnetism (superdiamagnetism) at a pressure of a few kilobars and at temperatures as high as 150–200 K, in specially prepared copper chloride CuCl. (A weak magnetic field cannot penetrate an ideal superconductor; this property is known as the Meissner effect. We can say formally that in the case of the Meissner effect, as for that of ideal diamagnets, the magnetic susceptibility is $\chi_{\text{ideal}} = -1/4\pi$. For normal diamagnets $\chi \approx \approx -(10^{-4}\text{–}10^{-6})$. I apply the term superdiamagnetic to those materials for which χ is comparable to χ_{ideal}, say, $\chi \sim -(0.01/4\pi\text{–}0.1/4\pi)$. It is clear from this discussion that superconductors are superdiamagnets, but the contrary is not necessarily true.) Unfortunately, it is still not clear whether this is a genuinely new effect, the result of an experimental error, or an imitation of true superdiamagnetism. If superdiamagnetism indeed occurs in CuCl it may be due to the high-temperature superconducting phase which can, in principle, appear when some semiconductors or semimetals switch over to the superconducting state (see[8], Chapter 5). Another possible cause is the formation of sandwiches of Cu and CuCl. There is, however, a quite different suggestion[13] that compounds of an unknown type can apparently exist which have spontaneous currents, and should be superdiamagnetic but are different from conventional superconductors.

The evaluation of the experimental results for CuCl and the theory for non-superconducting superdiamagnets are as yet insufficiently clear to make it appropriate to discuss them in detail here. But even if the results for CuCl do not reveal high temperature superconductivity, or a new physical effect, this by no means refutes the possibility of the existence of high temperature superconductors. The problem remains, and attempts at solving it appear to be extremely fascinating.

*During the last few Years three events worthy of notice have occurred. Firstly, the organic crystal di-tetramethyl-tetraselena-fulvolene-hexafluoro-phosphate, $(TMTSF)_2PF_6$, has been found to be superconducting. It is true that this crystal has metallic conductivity at low temperatures, and superconductivity with $T_c \sim 1$ K only under a pressure of a few kilobars[157]. Nevertheless, this finding seems to open up a new class of metals and superconductors, as organic compounds may be varied comparatively easily. Moreover, there are reasons to expect comparatively high critical temperatures for organic compounds[8]. Of course, no definite predictions can be made here, but the discovery of a new class of superconductors with low critical temperatures is of sufficient interest by itself.

Secondly, although further studies of CuCl have not clarified the situation, new results[158] evidence high-temperature superdiamagnetism in this material under certain, not altogether clear, conditions[11,12]. I believe that the authors of one review[159] are right when they note that the difficulties encountered in the analysis of the behaviour of CuCl are fairly typical of materials which have properties that are difficult to control as, for example, some semiconductors. The results are influenced by impurities, various lattice defects and residual stresses. Therefore, high-temperature superconductivity or superdiamagnetism of CuCl cannot be ruled out, and I even believe that it is probable.

Thirdly, a strong diamagnetic effect similar to that described for CuCl was found in CdS crystals treated with pressure quenching at liquid nitrogen temperatures[160] ($T = 77.4$ K). The pressure of about 40 kbar was decreased at a rate exceeding 10^6 bar/s. The authors do not report any details of the technique used for the treatment of these samples (this is perhaps explained by the address of their laboratory, the United States Army Armament Research and Development Command, Large Calibre Weapons System). This result for CdS stimulates interest in CuCl and in the, still mysterious, mechanism of high temperature superdiamagnetism.

3. New substances (the creation of metallic hydrogen and some other materials)

There exist on the Earth a great number of different substances; some have been produced artificially, while others occur naturally: chemical compounds, alloys, solutions, polymers, etc. Generally speaking, the creation of new materials is a matter of chemistry or technology and cannot be classified as a physical problem. But this is not so in the case of truly unconventional (we might call them exotic) materials. Among these are the high-temperature superconductors mentioned above or, for instance, hypothetical crystals with closely packed lattices which (if they could be created) would possess extremely high mechanical and thermal parameters[(14)]. For example, a carbon material with a closely packed lattice ("superdiamond") would have a hardness (compression modulus) greater by an order of magnitude than that of ordinary diamond. Unfortunately, I am not acquainted with the current state of this problem and do not even know whether it can be accepted as a physical problem. However, there is one such "new" substance whose creation and study constitutes an important and interesting problem, and which, incidentally, has attracted a lot of attention in recent times. This substance is metallic hydrogen.

Under normal conditions (for example, under atmospheric pressure) hydrogen is known to be in the molecular state: it boils at $T_b = 20.3$ K and becomes solid at $T_m = 14$ K. The density of solid hydrogen is $\varrho = 0.076$ g cm^{-3} and it is a dielectric. However, under a pressure high enough to crush outer electron shells, any substance must undergo a transition to the metallic state. The density of metallic hydrogen can be estimated roughly by taking the distance between protons to be of the order of the Bohr radius $a_0 = \hbar^2/me^2 = 0.529\times 10^{-8}$ cm. Hence we have $\varrho \sim Ma_0^{-3} \sim 10$ g cm^{-3} (here M is the proton mass ($= 1.67\times 10^{-24}$ g)). A lower density value is given by quantitative, though unreliable, analysis: for instance, according to the calculations[(15)], molecular hydrogen is in thermodynamic equilibrium with metallic hydrogen at a pressure $p = 2.60$ Mbar when the density of metallic hydrogen is 1.15 g cm^{-3} (in this case the density of molecular hydrogen is 0,76 g cm^{-3}). According to the analysis[(16)], the pressure at equilibrium lies between 1 and 2.5 Mbar, the uncertainty being due to the lack of reliable data on the equation of state for the molecular phase. Metallic hydrogen may be superconducting, with a critical temperature as high as 100–300 K (for metallic hydrogen the Debye temperature is about 3×10^3 K and for $g < 1/2$ formula (3) yields $T_c \lesssim 500$ K).

The creation of metallic hydrogen, which may be regarded in some respects as the simplest metal, and the determination of its critical temperature, are not only of obvious physical interest but may also have significant astrophysical implications. Suffice it to say that large planets such as Jupiter and Saturn must contain a considerable proportion of metallic hydrogen[17]. The much more important consideration is that metallic hydrogen may prove to be stable (although its state will, of course, be metastable) even under zero or comparatively low pressure. There are well-known examples of such quite durable metastable modifications: for instance diamond, which has a higher free energy than graphite at low temperatures and pressures. As for metallic hydrogen, calculations[16, 18a] have shown that it will be stable in the absence of pressure, but it is unclear as to whether the lifetime of such a state can be sufficiently long.

The theoretical analysis[16] of the possible structure of metallic hydrogen has yielded fascinating and unexpected results, despite the questions of the stability and lifetime of the metastable state. According to these results, at zero pressure, metallic hydrogen must have a filamentous structure which lacks order along the filaments, that is, it has only two-dimensional periodicity (a triangular lattice is formed by the filaments in the plane perpendicular to them). Under pressure metallic hydrogen may become liquid, even before the equilibrium pressure has been reached (at the equilibrium pressure metallic hydrogen co-exists with molecular hydrogen). In this case solid molecular hydrogen will, obviously, be transformed under pressure into liquid metallic hydrogen. It is possible, however, that the liquid state can exist at pressures exceeding the equilibrium pressure. At the same time, other results[18b] have been reported and, in general, the structure of metallic hydrogen is still unclear.

New advances in the studies of metallic hydrogen can hardly be expected in the absence of experimental work, namely, attempts to create it. Also, another task is to determine the parameters of molecular hydrogen under high pressures, which are still unknown. It may be interesting to study various alloys of metallic hydrogen with heavier elements. The problem of metallic hydrogen (both light and heavy hydrogen, that is, deuterium) is one of the most challenging at present. If the problem is solved "successfully" and metallic hydrogen proves to be sufficiently stable (long lived) at low pressures and is even superconducting into the bargain, then the production and study of metallic hydrogen will become one of the primary tasks of macroscopic physics.

Unfortunately, attempts to obtain metallic hydrogen under quasi-equilibrium conditions necessitate the production of pressures exceeding 1.5–2 Mbar in a defined volume. The materials available, including diamond, cannot generally withstand such loads and, therefore it is very difficult to manufacture a pressure chamber for the compression of hydrogen. An interesting approach to overcoming this difficulty is to obtain superhigh pressures in the small contact region between a tapered punch and a flat anvil made of diamond, or based on diamond. This method was used in experiments[19a] in which there was some indication of the appearance of metallic hydrogen. Other methods have been suggested and used[19b], but on the whole the problem is far from solved, and nobody can predict when we will be able to obtain reliable evidence of the transition of hydrogen to the metallic state, let alone obtain a "lump" of metallic hydrogen.

The creation or use of materials with extraordinary properties is a favourite subject for science fiction authors. Apparently, for them, everything is possible. But even respectable scientific journals have been known to report the discoveries of quite unusual substances which could not later be confirmed (for instance, polymeric of superhigh density water). The reason is, on the one hand, that in many cases these new substances are produced in very small amounts, for very short time intervals (as for example, during an explosion) or under very high pressure, so that their properties are hard to ascertain. On the other hand, the authors are, of course, tempted to claim a great discovery. Such cases are instructive, in particular, as a reminder that any discovery should be comprehensively checked and rechecked before being finally accepted.

* The pressure at which metallic hydrogen occurs can be lowered when one adds impurities or creates electron-hole pairs[211].

4. Metallic exciton (electron-hole) liquid in semiconductors

When conduction electrons and holes exist in a semiconductor (generated, for instance, by light quanta) then at sufficiently low temperatures they should combine, giving rise to excitons—hydrogen-like atoms similar to positronium. In a first approximation, the bonding energy and radius of such excitons in the ground state are

$$E_{0,\mathrm{e}} \sim \frac{e^4 m_{\mathrm{eff}}}{2\varepsilon^2 \hbar^2} = \frac{E_0 m_{\mathrm{eff}}}{m\varepsilon^2} \quad \text{and} \quad a_{0,\mathrm{e}} \sim \frac{\hbar^2 \varepsilon}{m_{\mathrm{eff}} e^2} = \frac{a_0 \varepsilon m}{m_{\mathrm{eff}} \varepsilon} \tag{4}$$

where $E_0 = e^4 m/2\hbar^2$ and $a_0 = \hbar^2/me^2$ are the well-known Bohr expressions for the energy and radius of the hydrogen atom, m_{eff} is the effective mass of electrons and holes (here their masses are taken to be equal and anisotropy is ignored), and ε is the dielectric constant of the semiconductor.

As in some cases we have $\varepsilon \gtrsim 10$ and $m_{\mathrm{eff}} \lesssim 0.1$ m we obtain an exciton radius $a_{0,\mathrm{e}} \gtrsim 10^{-6}$ cm and an energy $E_{0,\mathrm{e}} \lesssim 10^{-2}$ eV ~ 100 K. The difference between these parameters and those of a hydrogen atom lies in the fact that here the Coulomb interaction force is ε times weaker, and the effective mass may be small compared with the mass m of the free electron. (For the cases in which we are interested, the exciton radius $a_{0,e} \gg a_0 \approx 5 \times 10^{-9}$ cm. and this very fact allows us, generally, to describe the interaction between an electron and a hole by Coulomb's law, taking into account the effect of the medium; the interaction between charges $-e$ and $+e$ in this case is known to be attractive with energy $e^2/\varepsilon r$ where r is the distance between the charges.)

We have already mentioned in our discussion of metallic hydrogen that equality of the electron shell size and the distance between the nuclei can be taken as a rough criterion for high density and metallization of a material. For excitons in a semiconductor this means that their density is high for concentrations of $n_e \sim a_{0,e}^{-3} \sim 10^{18}$ cm^{-3}. Thus we see that, while a high hydrogen density can only be obtained at a pressure of millions of atmospheres, a high density of excitons corresponds to a quite normal concentration, $n \sim 10^{18}$ cm^{-3} of electrons and holes in semiconductors. Just this sole possibility of simulating ultrahigh pressures in semiconductors gives sufficient importance to the problem under consideration. This conclusion is strengthened if we consider the possible behaviour of a high density system of exitons in a semiconductor[20]. Such a system must go over to the liquid state and give rise to drops. Such drops most probably consist of an electron-hole metal that is, they are similar to a liquid metal. But it cannot be ruled out that these drops have a "molecular" structure; in this case they are similar to liquid hydrogen, consisting of H_2 molecules (the part of molecules in the molecular, and hence dielectric, exciton "liquid" is played by bi-excitons, that is, two excitons bound together). The electron-hole (exciton) liquid can, in principle, exhibit superconductivity or superfluidity. In short, the exciton liquid in semiconductors must possess various fascinating properties and peculiarities, depending of course, on the parameters of the semiconductor "container" being used. Extensive experimental studies in this field are underway, but the situation on the whole is still unclear[20] (however, see below). It is reasonable to think that exciton liquids in semiconductors will be one of the central problems of solid state physics in the near future.

Two more arguments could be presented to support this conclusion about the importance of the problem. Firstly, the system of excitons in a semiconductor can be used to simulate not only ultrahigh densities (pressures) but also the effect of ultrahigh magnetic fields. This will be discussed in Section 6. Secondly, next in line are studies of excitons in one-dimensional and two-dimensional systems, that is, on solid surfaces (two-dimensional or quasi-two-dimensional systems) and in various quasi-one-dimensional systems (long polymers, intersections of crystal edges, and thin crystalline filaments known as whiskers, and dislocations). Generally speaking, electron-hole "atoms" can be formed in such systems too[21], and the criterion for high density in this case will be $n_e \sim a_{0,e}^{-2}$ (for two-dimensional systems) or $n_e' \sim a_{0,e}^{-1}$ (one-dimensional systems). This implies that in a two-dimensional system the "liquid" transition occurs for exciton concentrations as low as $n_e \sim 10^{-12}$ cm^{-3} (for $a_{0,e} \sim 10^{-6}$ cm). Moreover, there are other highly interesting aspects of the problem of surface excitons, as for instance, those linked to the problem of high-temperature superconductivity[7,8]) (see also Section 8).

We have already stressed that the list of important and interesting problems discussed in this book is far from being complete and that the classification into the "important" and "unimportant" categories is a difficult, and above all, arbitrary procedure, We mention it here once more because we have selected for discussion only the problem of exciton liquids among all the problems of semiconductor physics. Physicists working with semiconductors would, undoubtedly add a number of other problems, such as metal-dielectric phase transitions, disordered systems, conductive polymers, etc. But I would still say that the problem of exciton liquids in semiconductors is distinguished by its novelty, elegance and complexity.

Up to this point the content of this section follows the text of the previous edition of this book (1974). I have done this intentionally to show the advances of the last years. Indeed, during this period the problem of exciton liquids in semiconductors has attracted considerable attention and has largely been solved[22]. The metallic exciton liquid has been shown to appear in germanium and silicon under certain conditions, and a number of experiments have been carried out with this amazing liquid. It would be too early, however, to strike the problem of exciton liquids from our list of important problems. We know that semiconductors have very different properties; for example they differ in their anisotropy. Accordingly, exciton liquids in semiconductors must have different properties and parameters. It would be especially interesting to analyse the possibility of the existence of a dielectric exciton liquid which has not yet been found in experiments, in contrast to metallic exciton liquid. As mentioned above, the next questions to answer are those about the superconductivity and superfluidity of exciton liquids. The other substantially unclear questions are related to the behaviour of exciton liquids in magnetic fields, in films, on surfaces and in laminar compounds[21,23].

As I noted above, there is no doubt that studies of exciton liquids in semiconductors will attract considerable attention in coming years. However, much has been done in this field and if I wrote the book anew I would, perhaps, discuss other problems of semiconductor physics and, in general, solid state physics. Apart from the problems mentioned briefly above, we should give special mention to the so-called spin glasses—a name that does not sound too attractive to a physicist. However, this is a very interesting and, to a certain extent, new class of materials. [In a spin glass the spins are randomly distributed in space, but their directions are ordered (correlated) in a certain, non-trivial, way. Therefore the term "spin glass" is a somewhat misleading one, especially as quite different metallic, and generally conducting, glasses exist, which differ from conventional dielectric glasses mainly in their conductivity, and which are now studied extensively(25)]. Nevertheless, I have decided to cite a few references on spin glasses(24).

5. Second-order phase transitions (critical phenomena)

The superconducting transition, the conversion of helium I into superfluid helium II, the transformation of a paramagnetic state into a ferromagnetic one, many ferroelectric transitions and some other transformations of alloys—these are all well-known examples of second-order phase transitions. During these transitions, no evolution (or absorption) of latent heat occurs, no discontinuity of volume or lattice parameters is found, and the transformation may be regarded, in a sense, as a continuous one. At the same time, discontinuities of specific heat, compressibility and other characteristics are observed at the transition point, and in its vicinity many of these characteristics behave anomalously. For instance, in the vicinity of the helium I $\rightleftharpoons$ helium II transition and some other transitions the specific heat may be satisfactorily approximated by $C \sim \ln |T-T_c|$ where T_c is the transition temperature (in this case, the lambda-point temperature). During ferromagnetic and ferroelectric transitions the magnetic permeability and the dielectric constant respectively tend to infinity when $T \to T_c$ and can often be approximated by the Curie law $\chi \sim |T-T_c|^{-1}$.

Some of the first-order transitions close to the critical Curie point (often referred to as the tricritical point) in the p, T diagram are similar to second-order phase transitions(26). The essence of the matter is that the variation of some parameters (for instance, pressure) may transform second-order transitions into first-order ones (the tricritical point is just the point of contact of the curves for transitions of these types in the p, T diagram). The first-order transitions close to the tricritical point are, naturally, akin to second-order transitions (the latent heat of transition is non-zero, but small, the behaviour of the specific heat is anomalous, and so on). Examples of such transitions are some ferroelectric transformations, the superfluid transition in mixtures of ^{4}He with ^{3}He, and, apparently, the $\alpha \rightleftharpoons \beta$ transition in quartz. Finally, second-order transitions are similar to critical points for liquid-vapour (gas), and some other critical points.

In order to solve the problem of second-order phase transitions (and transitions close to them[27]) we must obtain sufficiently full knowledge, both quantitative and qualitative, of various phenomena in the vicinity of the transition points. In particular, we have to find the temperature dependence of all parameters, that is, the dependence on the temperature difference $T-T_c$.

As second-order phase transitions are continuous it is natural to treat them by expanding thermodynamic functions (for instance, a thermodynamic potential) in powers of a certain parameter η which vanishes at equilibrium when $T \geqslant T_c$. Then coefficients, A, B, C, etc. in the resulting expansion

$$\Phi = \Phi_0 + A\eta^2 + B\eta^4 + C\eta^6 + \ldots \tag{5}$$

are, in their turn, expanded in powers of $T-T_c$ so that in the vicinity of a typical second-order transition we have $A = A'(T-T_c)$ and $B = B_0 = \text{const.}$ Landau[26] developed the theory based on the consistent use of this approach which can be traced back to Gibbs, van der Waals and others.

The Landau theory yields the Curie law $\chi \sim |T-T_c|^{-1}$ for susceptibilities, and the spontaneous magnetization $\mathcal{M}$ or the spontaneous polarization $\mathcal{P}$ at $T < T_c$ varies as $\mathcal{M} \sim (T_c-T)^{1/2}$ or $\mathcal{P} \sim (T_c-T)^{1/2}$ and so on. At the same time, the Landau theory is, generally, incapable of explaining the anomalous temperature dependence of the specific heat and other parameters at $T \to T_c$. Moreover, careful measurements[26, 28] have revealed that the Curie law and similar relationships are not valid in the immediate vicinity of the transition point where $\chi \sim |T-T_c|^{-\gamma}$ and $\mathcal{M} \sim (T_c-T)^{\beta}$ but $\gamma \neq 1$ and $\beta \neq 1/2$.

The Landau theory yields the same results as model theories (such as the well-known Weiss theory of ferromagnetism) which are based on the method of self-consistent (sometimes called molecular) fields. This fact indicates what is in itself basically clear, namely that the limitations of the Landau theory lie in ignoring fluctuations. Indeed, the theory deals with a mean value, for instance, of magnetization. But when $T \to T_c$ the mean value $\mathcal{M} \to 0$ while the fluctuations of $\mathcal{M}$ not only do not vanish, but on the contrary, grow where fluctuations are relatively small and this range is different for different sharply. Thus, it is clear that the range of validity of the Landau theory is one transitions[26, 29]. But, near the transition (that is, where the difference $|T-T_c|$ is sufficiently small) we must take into account fluctuations, thus obtaining anomalous specific heat behaviour, deviations from the Curie law $\chi \sim |T-T_c|^{-1}$ and so on.

No consistent theory of second-order phase transitions in three-dimensional systems has yet been fully developed although strenuous efforts have been made to solve this problem. (L. D. Landau once told me that his attempts to solve the problem of the second-order phase transition had demanded greater effort than any other problem he had worked on). But these efforts have by no means been wasted, and in recent years important advances have been made. These include, first of all, the similarity laws[26,28] which give the temperature dependence of various parameters in the vicinity of the transition temperature T_c. Using these laws and some experimental data, we can, for instance, predict that in some cases when $T \to T_c$ the magnetic susceptibility $\chi \sim |T-T_c|^{-\gamma}$ where $\gamma = 4/3$ (rather than $\gamma = 1$ as given by the theories of Weiss or Landau). Furthermore, the so-called critical indices (β, γ, etc.) can be calculated fairly accurately for various systems without using experimental data. [Interestingly, the fact that a self-consistent theory of the type of the van der Waals theory (or Landau theory as it is often termed now) is inapplicable in the vicinity of the critical point in a liquid, was noted as early as the end of the last century[30]].

Thus, one of the central problems of solid state physics is to develop a consistent theory of second-order phase transitions and related transitions, which accounts for typical features of various transformations, as well as to give a general description of the kinetic processes in the vicinity of T_c. Although some people believe that the problem has been basically solved this in any case cannot be said about some interesting special cases.

This can be illustrated by two more specific problems in this area whose selection is, perhaps, arbitrary as they lie within my own sphere of interests. The first is the behaviour of helium II in the vicinity of the lambda point. In the Landau theory of superfluidity the density ρ_s of the superfluid helium component is taken to be a specified function of, say, the temperature T and the pressure p. But according to the general theory of second-order phase transitions, the density ρ_s cannot be specified, but must be determined from the condition of the minimum of a thermodynamic potential. This approach[31] yields a number of interesting results: the dependence of $T_c \equiv T_\lambda$ and specific heat C on the thickness of the film helium II; non-uniformity of the density ρ_s in the vicinity of a solid wall or a vortex axis in helium II, and so on. Apparently these results reflect reality, but, on the whole, the theory of the superfluidity of helium II in the vicinity of the lambda point and its experimental verification, are far from being complete.

The second problem is the scattering of light around second-order phase transitions and particularly, near the $\alpha \rightleftarrows \beta$ transition in quartz[32–35]. Since fluctuations grow as the temperature approaches T_c we can expect an increase in the intensity of scattered X-rays, neutrons or light in this region. A similar phenomenon (critical opalescence) has long been known to occur near the liquid-vapour critical point. A sharp increase in the intensity of scattered light is also observed in quartz[32] around the transition from the α modification to the β modification which occurs at a temperature

$T_c = 846$ K. The picture had seemed to be basically clear here but it was later shown to be more complicated[34] and cannot be described in terms of a simple theory[33].

It does not seem too strange now since the theory[33] ignored "for the sake of simplicity" such a fundamental difference between solids and liquids as the possible existence of shearing stresses in solids. Indeed, isotropic solids, for instance glasses and even crystals, exhibit almost the same behaviour as liquids in some cases (or, more exactly, when certain effects are studied in them). But this approach of course, cannot be applied indiscriminately. For instance, in liquids transverse sound waves are strongly damped and can be said not to propagate, while in solids the propagation conditions for transverse sound waves are, generally, no worse than for longitudinal sound waves.

It has been shown[35] that shearing stresses must be taken into consideration in the analysis of the scattering of light in solids and, particularly, the anomalous scattering near phase transitions. Some features of light scattering in crystals can be explained by the inclusion of this factor, but this does not give a straightforward explanation of the scattering picture in the case of the $\alpha \rightleftarrows \beta$ transition in quartz. In this case (and, apparently, in some similar cases) a complicating factor is the appearance of an inhomogeneous phase or a mixture of α and β phases in the region of the transition. The problem of light scattering in the vicinity of the $\alpha \rightleftarrows \beta$ transition is still, on the whole, unclear, both theoretically and experimentally. This stresses the need for further studies of the scattering of light near phase transitions in solids, liquid crystals and liquids[35] (in particular, in liquid helium). There can be no doubt about the potential fruitfulness of this work, particularly, if one recalls the important results obtained in studies of light scattering in liquids and solids outside the regions of phase transitions[36].

It should be noted that the problems discussed above deal with what might be called everyday or classical phase transitions. Recently, physicists have become increasingly interested in such exotic transitions as the transition of liquid ^{3}He to a superfluid state[37], phase transitions in the exciton "matter" in semiconductors (see Section 4), the transition to a superfluid in liquid molecular hydrogen[38a] and in quantum crystals[38b], phase transitions in substances of ultrahigh density, in particular, in neutron stars (see Section 21), and so on.

We should also mention phase transitions in non-quantum liquids: in liquid crystals, in magnetic materials (ferromagnetic and antiferromagnetic transitions in the liquid phase), the ferroelectric transition in liquids; phase transitions in solids giving rise to "incommensurate" (inhomogeneous) phases, and finally phase transitions and various anomalies (such as, for example, an alteration of the temperature dependence of the magnetic susceptibility occurring at a certain point T_a) in quasi-two-dimensional and quasi-one-dimensional systems (see Ref. 8, Sections 6 and 7).

From among the transitions mentioned above, I have selected for discussion here only the transitions of liquid ^{3}He. About 20 years ago, a suggestion was made that the atoms of ^{3}He could "stick" to each other, forming pairs with integral spin, and could go over to a superfluid state due to Bose–Einstein condensation (see Section 2). This state is similar to a superconducting state but, as ^{3}He atoms are neutral, it must be superfluid rather than superconducting. Incidentally, superconductivity may also be referred to as superfluidity, but in a system of charged particles.

It had been thought that ^{3}He atoms paired up under the action of rather weak van der Waals forces, so that the temperature of the superfluid transition must be extremely low. But, in 1972 and 1973, it was found that not one, but two phase transitions[37] occur in liquid ^{3}He at the very low but still attainable temperatures of about 2.6×10^{-3} and 2.0×10^{-3} K, respectively, though at a pressure of about 34 atm. Now it is known for sure that these are, indeed, two transitions to the superfluid state which differ from each other in the total angular momentum of the pairs. The attractive force responsible for the formation of pairs seems to be due, not to the van der Waals interaction, but mainly to the interaction between the spins of ^{3}He atoms (this is the so called exchange interaction which gives rise to ferromagnetism).

Astonishingly good extensive studies of superfluidity and other effects in liquid ^{3}He have been carried out in the last few years. Incidentally, the abundance of ^{4}He is about 10^7 that of ^{3}He. The relevant experimental techniques must be extremely good since superfluid ^{3}He is much more difficult to work with than superfluid ^{4}He because of the existence of the orbital angular momentum and spin angular momentum and because the experimental temperature range is just $(1–3)\times10^{-3}$ K above absolute zero. In my opinion, the most impressive advances made in the physics of condensed media in recent years are those of the studies of liquid ^{3}He.

Intensive studies of liquid ^{3}He and transitions in the systems mentioned above continue. The problem of phase transitions as a whole remains one of the central fields in macrophysics.

* Recently published results fully confirm the opinion voiced above. However, attention is drawn increasingly to new systems and phenomena, some of which have already been mentioned—for instance, transitions in liquid crystals and transitions giving rise to "incommensurate" phases. They also include possible transitions in DNA and in polymers, and the superfluid transition in the gas of atomic hydrogen. This gas, if it can be created, will be rapidly transformed, into a gas of molecular hydrogen (H_2)

under normal conditions. But at a low temperature $T \lesssim 1$ K the gas of atomic hydrogen can exist for many minutes in a vessel whose walls are coated with superliquid helium II[161]. Moreover, if a sufficiently strong magnetic field is applied to the gas the stability of atomic hydrogen increases[162] and, apparently, conditions can be created under which recombination can be ignored. The reason for this is well known, i. e. that in the H_2 molecule the directions of electron spin oppose each other, while in a strong magnetic field all electron spins have the same direction. In order to form a H_2 molecule the electron spin of one of the H atoms must flip, which is not easy. H atoms in the ground state with parallel spins repulse each other (to be more precise, there is a weak van der Waals attraction between atoms only at long range). Therefore, such a gas cannot be liquefied at atmospheric pressure even at temperatures down to absolute zero. But the Bose gas of hydrogen atoms must undergo Bose–Einstein condensation at a temperature which depends on the gas density, and the resulting phase must be superfluid (it is important here that the gas is not ideal, and the inclusion of the interaction leads precisely to superfluidity).

Studies of phase transitions in two-dimensional and quasi-two-dimensional systems are particularly impressive in their scope and significance. Strictly speaking, the problem is not exactly new[27] but its importance is steadily increasing[163–165]. A wide variety of the tasks in this field are, naturally, closely related to surface physics (see Section 8).

6. Matter in ultrahigh magnetic fields

The characteristic difference between the energy levels of the hydrogen atom is

$$E_a \sim \frac{e^4 m}{2\hbar^2} \sim 10 \text{ eV} \tag{6}$$

The energy difference between the levels of a non-relativistic free electron in a magnetic field is

$$E_H \sim \frac{e\hbar H}{mc} \sim 10^{-8}\, H \text{ eV} \tag{7}$$

where the magnetic field is measured in Oersted (or in Gauss since H may be replaced by the magnetic induction B). The estimate given by eq. (7) is valid both for orbital levels and for spin levels (the electron magnetic moment (μ) is $e\hbar/2mc$, and the energy difference between the levels of states with a spin parallel and antiparallel to the field is precisely $2\mu H = e\hbar H/mc$).

Until recently, scientists worked with magnetic fields which were weak on the atomic scale, that is, $E_H \ll E_a$ and hence

$$H \ll \frac{e^3m^2c}{\hbar^3} = \left(\frac{e^2}{\hbar c}\right)^2 \frac{mc}{e\hbar} mc^2 \sim 3\times 10^9 \text{ Oe}. \tag{8}$$

For heavy atoms (atomic number Z), a similar criterion for a weak field is $H \ll 3\times 10^9 Z^3$ Oe, that is, enormously high fields are needed to overcome it. Therefore, even relatively recently, the question of high fields

$$H \gtrsim 3\times 10^9 Z^3 \text{ Oe} \tag{9}$$

for which was regarded as a rather academic one, and did not attract any special attention. But now the situation has changed.

The pulsars discovered in 1967–1968 are magnetic neutron stars whose surface magnetic fields are estimated to be as high as 10^{13} Oe (see Section 21). This implies that material on and close to the pulsar surface is in a high field, even allowing, for the fact that this material is mainly iron ($Z = 26$, $Z^3 \sim 20\,000$). In high fields, for which condition (9) is satisfied, and especially in the ultrahigh fields for which $H \gg 3\times 10^9 Z^3$ Oe, atoms are not seen in the conventional forms of atoms in weak fields or in the absence of fields. In ultrahigh fields the electronic shell of an atom is stretched into a relatively thin needle along the field. Under such conditions it is energetically favourable for two atoms—say, for example, of iron—to form a molecule (Fe_2) with a high bonding energy. A set of such molecules would, probably, give rise to a polymer-type structure also with a high bonding energy (for a discussion of the behaviour of matter in high magnetic fields, see Ref. 39). This is precisely the structure that the solid surface of neutron stars would have if that surface field were high enough. As mentioned above, this seems in some cases to be quite probable. It would be very hard to vaporize or destroy this surface crust with an electric field and this may have significant implications for the theory of pulsars[39a, c].

Pulsars are at a great distance from us, and this, of course, makes it very difficult to investigate the behaviour of matter in ultrahigh magnetic fields. Anyway, most physicists prefer to deal with "terrestial"conditions and are little excited by the possibilities of astrophysical studies. Quite apart from these considerations, it is natural to ask whether high fields can be generated and utilized in a laboratory environment. The prospect here does not seem to be very promising, even if we take into consideration the fields produced by superpowerful lasers (see Sections 7 and 15). Quite different prospects for studying the effects of ultrahigh fields are opened up by their simulation in exciton systems of semiconductors. Indeed, as shown by eq. (4) the bonding energy of the hydrogen-like exciton is smaller than that of the hydrogen atom by a factor of $m_{eff}/m\varepsilon^2$ that is, for instance, under practical conditions by a factor of 1000. The energy of electrons and holes related to their orbital motion in a magnetic field is given by eq. (7), where m is replaced with m_{eff}. This implies that the magnetic field acting on an exciton is high, starting from fields as low as $H \gtrsim 3\times10^9\, m_{eff}^2/m^2\varepsilon^2 \sim 3\times10^5$ Oe (when $m_{eff} \sim 0.1$ m and $\varepsilon \sim 10$) and sometimes even lower[20,22,166]. Such fields can be generated, although usually as pulses. This will, probably, make it possible to study the exciton "matter" in high and even ultrahigh fields.

My classification of the problem of matter in high magnetic fields as an "especially important" problem may give rise to objections and doubts, as in some other cases. But I believe that this problem is distinguished by its refreshing novelty and the unexpected implications concerning neutron stars and excitons.

7. Rasers (*X*-ray "lasers"), grasers and superpowerful lasers

Lasers undoubtedly make a great contribution to the development of science and technology, although enthusiasm for lasers seems sometimes to be a matter of fashion. It has even been declared, albeit facetiously, that the atomic age has ended and the laser age begun. However, as it has already been explained in the introduction, in this small book we cannot discuss the development of laser technology or questions related to laser applications (including self-focusing and other non-linear processes and phenomena, which are, of course, very interesting in many respects). But the devices discussed in this section are exceptions which can hardly be left out of any list of "especially important" physical problems. It is, incidentally, difficult to understand why the first laser was built in 1960 and not about 40 years earlier, when Einstein put forward the clear-cut concept of stimulated emission which provided the basis for modern lasers. Apparently, the reason for this was that for a long time it seemed in principle to be only possible to amplify radiation by stimulated emission. The gain produced in this way is usually small, and so the impetus to laser development was given by the replacement of the amplification mode of operation with the mode of generation using multiple reflection of the light beam from mirrors confining the active medium in the laser. The concept of generation was in itself an innovation that came more naturally from physicists working in radiophysics, than from those working in optics.

To be fair, the above remarks are somewhat pedantic. It is often that an important discovery or innovation seems long overdue. Two vivid examples are Cerenkov radiation and the Mössbauer effect. And, of course, no "belatedness" of a discovery or an innovation can belittle the credit going to those who finally made it.

Modern lasers have a pulse power of up to 10^{13} W. The limit of focusing is, in fact, given by an area λ^2 ($\sim 10^{-8}$ cm^2), where λ($\sim 10^{-4}$ cm) is the typical wavelength of existing high-power lasers. Thus, the energy flux density would be as high as $P \sim 10^{21}$ W cm^{-2}. The smallest focal area so far obtained for a high-power laser is about 10^{-4}–10^{-5} cm^2, so that $P \sim 10^{17}$–10^{18} W cm^{-2}. In this case the electric (and magnetic) field strength at the focus is $\epsilon \sim 3\times 10^7$ esu cm^{-1} $\approx 10^{10}$ V cm^{-1} (this follows from the expression $P = c\epsilon^2/4\pi$ for the energy flux in an electromagnetic field). As was mentioned in Section 1, the laser power must be increased by two orders of magnitude to build laser fusion reactors, not to mention the efficiency, which must also be increased by an order of magnitude or so. This task is very complicated and cannot be regarded as a purely technical one; its fulfilment depends on research results. And it is a purely physical problem to raise the energy flux density P up to the 10^{26}–10^{27} W cm^{-2} needed for the efficient production of electron–positron pairs *in vacuo* (see Section 15).

Another fundamental physical problem is the development of the X-ray and gamma ray analogues of the laser which are often referred to as rasers and grasers (or gasers), respectively. The word "laser" is the acronym for "light amplification by stimulated emission of radiation". Therefore, the often-used terms "X-ray laser" and "gamma laser" are seemingly contradictory. The words "raser" and "gaser" are produced from "laser" by replacing the letter l (light) with the letters r (Roentgen) and g (gamma). One reason for this etymological discourse seems to be the lack, on my part, of more tangible subject matter; that is, any specific suggestion for the design of rasers and gasers.

On the subject of terminology, we should remember that no distinct boundary exists between X-rays and gamma rays. Sometimes the upper limit of the photon energy of X-rays is taken to be 100 keV, which corresponds to a wavelength of about 0.1 Å. Then gamma rays consist of photons of higher energies. No less often, radiation is classified according to its origin; nuclear transitions are assumed always to produce gamma radiation. All the reported concepts of rasers and gasers, that I know of, deal with photon energies of not more than about 10 keV so that gamma rays here are the photons emitted in nuclear transitions.

The development of rasers and grasers will be possible only when enormous difficulties are overcome. Firstly, the gain of the device falls rather rapidly with decreasing wavelength, so that, in general, an extremely high pumping power is needed to provide the required inverted population of excited levels (it will be recalled that stimulated emission usually comes from transitions from excited atomic or nuclear levels to lower ones). Secondly, in the X-ray range, and no less in the gamma ray range, it is very difficult to build a good resonator (reflectors) that will allow the radiation to remain for a long time in the excited active medium. Of course, in lasers this is done by mirrors set up at the ends of the active medium. In principle, we can dispense with reflectors, but then either the gain must be high or the size of the system—the active medium—large. Therefore, raser design may involve, for instance, amplification of radiation in high-density relativistic electron beams which can, in principle, be produced by high-current accelerators.

A possible concept for a graser design is to use Mössbauer-type nuclear transitions (very narrow lines are needed) in systems in which the upper level is populated because of the capture of neutrons produced by nuclear explosion (the "neutron pumping" power must be high). But other practical ways for solving the problem are being sought[(40)].

The obstacles to the development of rasers and grasers are so great that one may ask whether such devices, which may be useful for physics, can be built at all. [We are discussing here devices emitting radiation with wavelengths of an ångström or shorter; but it is, apparently, much easier to build a raser with longer wavelengths]. The fact is that the successful development of lasers was made possible not only by the fundamental principles, (there was no doubt about the feasibility of stimulated emission of radiation in all spectral ranges), but also by the widest choice of optical techniques, in combination with the relatively modest demands for pumping power and the properties of the active medium.

But who knows? The history of physics presents many examples of problems for which the hope of solution seemed dim or almost fantastic. And then new phenomena were discovered (for example, the fission of heavy nuclei) and what had been hopeless became practicable, and then even common. Maybe we may expect some breakthrough in the development of rasers and grasers, such as the introduction of a new concept or the discovery of a new phenomenon. However, there is no guarantee of a "happy ending". It may be that somebody will find convincing evidence for the impossibility of the development of efficient and useful rasers and/or gasers.

* The work on lasers is, of course, continuing. But, apart from unconfirmed rumours, I have not heard of any dramatic developments in lasers or in gasers and rasers. Much has been written in recent years about free electron lasers[(167)]. These laser systems realize, and sometimes considerably modify, the fairly old idea of the generation of electromagnetic waves by a beam of relativistic electrons passing through an undulator or wiggler, which,

in its simplest form, is a system of magnets which produce a variable field which oscillates electrons in the beam. The analogy between such systems and lasers seems to be far-fetched, and the term “free electron laser” does not seem quite appropriate. Of course, the name does not matter. Hopefully, free electron lasers will prove to be practical and useful in optics and in microwave techniques. As for the X-ray spectral range, it is still quite unclear whether undulator-type systems using high-density relativistic electron beams (mentioned above in another context) can be effective.

It should be noted that high-power X-ray beams can be produced with synchrotrons (or with a linear electron accelerator in combination with the undulator) or by using plasma focus. In such systems we typically obtain incoherent radiation from individual electrons; but it is desirable that rasers, like lasers, produce coherent radiation, thus making possible a high degree of monochromaticity. In electron beam systems coherent radiation can be obtained only with high-density beams and under some other conditions which are difficult to satisfy in the X-ray range of wavelengths. This is precisely why it is unclear, as mentioned above, whether free electron lasers can operate in the X-ray range. Incidentally, the term “laser” is only applicable if we are dealing with coherent radiation.

8. Very large molecules, liquid crystals, some surface phenomena

The long title of this section covers three sets of problems which I would prefer to ignore, but which must at least be mentioned. The same may be said about an old or, rather, ancient, problem such as the origin of ball lightning. An exceptionally wide variety of explanations for this phenomenon have been put forward—including plasma formation, low-frequency and high-frequency discharge, antimatter, optical illusion and some physiological effects in the eye that results from the lightning[41, 42]. This demonstrates that the nature of ball lightning remains mysterious although recent studies[42] have, apparently, narrowed the range of possible explanations.

Very large molecules (nucleic acids and proteins) are prevalent in biology, and physical methods have proved indispensable in their study. But this is not enough; new and effective procedures must be developed for the analysis of the structure of large molecules, particularly in situations where their number is small, or they are in solution or mixture with other molecules[2]. This is, undoubtedly, a physical problem, and is both important and difficult.

Liquid crystals are those numerous substances which can simultaneously be in both a liquid and an anisotropic state (a substance in the liquid crystal state flows, but remains optically anisotropic). Liquid crystals have been studied for about 90 years but until recently they were regarded as something exotic. This is easy to understand. As long as the knowledge of simple substances (solids and liquids with simple structure, chemical composition, etc.) was insufficient, there was little hope of explaining the properties of considerably more complex substances. Moreover, studies of liquid crystals have not been stimulated by any prospect of important scientific or technological applications. But now the situation has changed dramatically. The "simpler" aspects of the physics of solids and, to a lesser degree, of liquids, are now fairly well understood, and those who insist on studying the simplest objects and processes appear increasingly like the man in the well-known joke who looks for his lost key under a street lamp, for the perfectly good reason that there is more light there. Moreover, liquid crystals have not only proved to be important for biology, but may be useful in some significant technological applications because of the strong dependence of some of their properties on temperature and external electric and magnetic fields[(43)].

Of special interest is the study of the different phase transitions in liquid crystals particularly with special reference to the general theory of phase transitions (see Section 5). This has led to a sharp increase in the number of papers on liquid crystals published in the most widely read and "influential" physical journals, in addition to publications on physical chemistry and the appearance of the specialized journal *Molecular Crystals and Liquid Crystals*, the publication of which is, in itself, a sign of growing interest.

Scientists have for a long time been interested in surface effects and the properties of solid and liquid surfaces. But only recently has it been possible to obtain really clean and perfect surfaces and to control and study them in detail[(44a)]. The present demands of miniaturization in technology and engineering put special emphasis on studies of surface phenomena and their practical usage. Very soon, by using the finest modern techniques, we may expect a sharp increase in advanced surface studies. Physicists will, apparently, intensify their studies of various surface elementary excitations, such as phonons, plasmons, excitons, spin waves, etc., localized near the surface[(44b)]. They will also search for surface analogues of ferromagnetism, ferroelectricity, superconductivity, superfluidity, and the liquid crystal state. Actually, in two-dimensional systems some phase transitions, in particular superconducting and superfluid transitions, differ in character from their respective transitions in three-dimensional systems. For instance, in a two-dimensional system, superconducting or superfluid and, in some cases, ferromagnetic long-range order cannot appear: for example, the magnetic moments are not parallel over any distance. But for finite, yet sufficiently large, surfaces, ordering is possible. Moreover, superconductivity and superfluidity may occur even in the absence of long-range order[27, 163–165].

Problems of order and various associated phenomena in two-dimensional and one-dimensional systems (polymer chains, crystal edges, etc.) should give rise to numerous studies both of the general theory and of various particular cases or specific conditions. Here one must not think that two-dimensional or one-dimensional systems are literally films or chains with a thickness of one atom or molecule (though even under such conditions the thickness of a film or chain is, of course, not zero but about 10^{-8}–10^{-7} m or more for larger molecules). Indeed, films which are a few atoms thick (for instance, a film of a few atomic layers deposited on a "substrate"), or even thicker objects possessing an appropriate lamellar or filamentous structure, exhibit certain properties typical of two- or one-dimensional systems and are often referred to as quasi-one- or two-dimensional systems. Of course, the above discussion is to a certain extent applicable to such systems too, and in general, also to clean surfaces of solid massive bodies, various films (including surface films), chains, edges, and so on.

The three problems discussed in this section are treated somewhat vaguely in comparison with other problems mentioned in the book. Perhaps I have failed to produce sufficiently convincing arguments and evidence to support and substantiate my opinion that studies of large molecules, liquid crystals and some surface phenomena are now of especial importance. It can only be added that all the above problems with the possible exception of liquid crystals, involve searching for essentially new methods or effects, and expanding the scope of macroscopic physics.

* I got better acquainted with the new and exciting work in surface physics when I took part in the Summer School[165] in Varna, Bulgaria, in October 1980. A wide variety of techniques are used in this field, such as low energy electron diffraction (LEED), angle resolved photo-emission spectroscopy (ARPS), the inelastic scattering of ions with energy of about 1 MeV, electron microscopy, the study of surface acoustic waves and surface polaritons, and so on. These techniques make it possible to study both clean surfaces and individual atoms or molecules at the surface. Particularly interesting results have come from the study of inversion layers at the Si–SiO_2 boundary, properties of electrons at the surface of liquid helium, and the reconstruction of some crystalline surfaces[165, 168].

The reconstruction of the surface means here the variation of the lattice parameter for atoms at the surface. For instance, at the silicon surface (the crystal face 111), under certain conditions, the lattice parameter is seven times that in the bulk of the crystal. The analysis of reconstruction effects in some cases may take into account the contribution of the surface electron levels. The atoms on the surface, and even in the neighbouring layers, are clearly under different conditions from those of atoms in the bulk of the crystal. It should not be surprising, therefore, that the crystal lattice or, say, the magnetic phase at the surface, differs from that in the bulk of the crystal. Thus, wide-ranging prospects for new possible effects are opened. Obviously, surface physics is in a rapid growth period now, and many new and significant results are forthcoming.

9. Superheavy elements (far transuranic elements). Exotic nuclei

The heaviest naturally occurring element—uranium—contains $Z = 92$ protons and $N = 146$ neutrons in its nucleus (^{238}U). Since 1940 scientists have produced artificial transuranic elements by bombarding heavy nuclei (including the nuclei of uranium and transuranic elements) with neutrons and various nuclei. The first artificial transuranic element was neptunium Np_{93}. This was followed by plutonium Pu_{94}, americium Am_{95}, curium Cm_{96}, berkelium Bk_{97}, californium Cf_{98}, einsteinium Es_{99}, fermium Fm_{100}, mendelevium Md_{101} and the elements 102, 103, 104, 105, 106 and 107, which do not yet have official names. The heaviest of the known transuranic elements live for a few seconds or even fractions of a second (the nuclei decay owing to emission of alpha and beta particles and spontaneous fission). A rough extrapolation indicates that elements with $Z \gtrsim 108$–110 must decay spontaneously at such a high rate that the production and analysis of such elements can hardly be possible. However, the properties of transuranic elements do not change monotonically either with increasing Z or with the parameter Z^2/A (where $A = Z+N$ is the mass number), even though these elements contain 240–260 particles (nucleons), and resemble in this respect drops of a liquid. In other words, one-particle and shell effects are noticeable and sometimes significant even for the heaviest elements. This suggests the possible existence of relatively long-lived isotopes of elements with $Z > 110$. Specifically, it has been suggested[45, 46] that the element with $Z = 114$ has a closed shell (that is, 114 is a magic number) and the isotope $^{298}114$ of this element containing $N = 184$ neutrons has even "double" magic. This does not imply that the $^{298}114$ nucleus is the most stable one, since we must take into account all the possible pathways of decay (spontaneous fission, alpha and beta decay). Some calculations indicate that the nucleus $^{294}110$ should have the longest lifetime with a half-life $T_{1/2}$ of about 10^8 years.

It seems to be universally recognized that all such calculations are not accurate enough, and have no quantitative significance whatsoever. But the increased stability of nuclei in the vicinity of $Z = 114$ and $N = 184$ seems to be possible, and high stability cannot be ruled out for some, or at least one, of their isotopes. If there is such an isotope it can be found on Earth, in meteorites or in cosmic rays. Apart from that, it may be hoped that the methods that have been successfully applied to produce the known transuranic elements will be used to synthesize more or less stable isotopes (say, with $T_{1/2} \gtrsim 1$ s).

All these approaches have been used for years in the search for far transuranic elements[45, 46]. This search is of considerable interest for nuclear physics and, possibly, for astrophysics; and, of course, it is as fascinating as the search for unknown or extinct animal species.

So, the search for far transuranic elements is continuing, although so far unsuccessfully. True, there was a sensational report in the middle of 1976 about the discovery of stable elements with $Z = 116$, 126 and so on. But even though the work was done by skilled physicists and published in a prestigious journal (*Physical Review Letters*), it proved to be erroneous. I mention it here only to emphasize that only those who do not work, do not make mistakes. Moreover, there are no grounds for delaying publication of sensational reports until they have been verified. From the viewpoint of the development of science it is more beneficial to publish an erroneous report (thus making possible its prompt verification by other authors), than to delay publication of really significant results until they have been verified. Of course, I do not call for the lowering of standards and allowing the publication of "raw" material: I just believe that too high a requirement for publication and too severe criticism of mistakes are unjustified. One should bear in mind that the author of a published erroneous work is severely punished just by the disclosure of his mistakes.

Returning to the problem of superheavy elements, I believe that nobody can object to including this problem amongst the most important ones. It is quite another matter as to whether we can, or should, classify superheavy elements as a problem of macroscopic physics. It is, of course, a controversial question and I shall touch upon it in the next section. Of more importance is the fact that only superheavy elements are discussed in the book, although nuclear physics has other problems worthy of mention here; for instance, the isomerism of nuclei due to differences in their shape[47], or the problems of nuclei consisting of nucleons and antinucleons. The fact is that, along with "conventional" nuclei there exist, so to speak, "exotic" nuclei such as the nuclei of nucleons and antinucleons which have already been observed[48] and which can be of a great interest for the understanding of nuclear forces and other studies. In this connection I may mention here the hypothesis[49] that under certain conditions there can exist stable or metastable nuclei which have a higher density than normal atomic nuclei and whose other parameters are, of course, also different. In general, the atomic nucleus is a very peculiar system, particularly because even in the heaviest nuclei the number of particles is not very large (not more than 300). Therefore, surface effects play a significant part in nuclei and various fluctuations in the distribution of levels are observed.

Finally, knowledge of nuclear forces is insufficient and this constitutes an essential difference between nuclear physics and atomic physics.

* A report(169) appeared at the end of 1980 of a possible observation of the track of a nucleus with $Z \geqslant 100$. The track was found in an olivine crystal of meteoritic origin. The nuclei found in cosmic rays leave tracks in crystals which can be revealed after special treatment (in particular, etching and annealing). The track length depends on the atomic number of the nucleus. About 150 tracks of length 180–240 microns were found in this study, and were identified as tracks of nuclei of the uranium-group elements.

The track corresponding to the nucleus with $Z \geqslant 110$ had a length of 365 microns. Of course, this result has to be verified; other such tracks, and additional evidence must be found that the track indeed belongs to a nucleus with $Z \geqslant 100$. If this is really so then the abundance of such nuclei is about 10^{-3} of the abundance of the uranium nuclei.

As for other problems of nuclear physics, it should be noted that studies of nuclei in some cases yield information as to the nature of the interaction between nucleons and between nucleons and leptons(170). Considerable interest is generated by the problem of nuclear matter which exists primarily in neutron stars. This problem is clearly linked to astrophysics(171) (see also Section 21).

I have mentioned above the possible existence of nuclear matter whose density is twice (or more) that of conventional matter(49). Such a high-density phase, apparently, cannot exist for known nuclei, but it has been suggested that its precursors can be observed (phase transitions are known to affect the properties of a material even before the transition point has been reached(172)). In general, nuclear studies continue to be concerned with fundamental physical problems.

II

Microphysics

10. What is microphysics?

Our discussion of macrophysical problems did not require any special introduction. But before we start a discussion of microphysics we must define the term. By microscopic size we mean the size of an atom (about 10^{-8} cm) and, of course, an atomic nucleus (about 10^{-13}–10^{-12} cm), so that atomic and nuclear phenomena should be classified as microphysics. But actually things are not as simple as that.

The notions "large" and "small" in physics (and not only in physics) can, of course, be used only with reference to a quantity (standard) which is regarded in itself as being neither large nor small. A natural standard of length (distance in space) is a characteristic size of the human body, for instance, a metre. However, for example, as well as atoms and nuclei, optical wavelengths and even some man-made objects are very small in comparison with this standard. And yet one can hardly say that thin films or wires of about 1 micron diameter are microscopic objects. One may add here that, compared to a metre, the Earth's size, or the distance between Earth and Sun (1.5×10^{13} cm) are very large. Thus, there would be no less reason for the Solar System, than for atoms and atomic nuclei, to be distinguished from macroscopic objects of the order of metres, if it were only a comparison of scales that was involved.

For these, and similar considerations, microphysics is often defined as the domain of quantum laws, while macrophysics is defined as the domain of classical laws. This approach seems to be quite justified, although its conditional character is clear. It should be said that the classical equations are sometimes applicable to collisions between nucleons, while the behaviour of purely macroscopic systems is sometimes governed by quantum mechanisms (as illustrated by the quantization of the magnetic flux in superconducting cylinders). Finally, one must bear in mind that the development of science is generally accompained by shifts of the boundaries between various fields and subjects, and with changes in concepts and approaches.

The above discussion suggests that we can regard the boundary between microphysics and macrophysics as a historically variable concept. Indeed, it seems reasonable to assume that now atomic and nuclear physics belong essentially to macrophysics rather than microphysics.

The reasons for this are as follows: Firstly, atoms and nuclei are systems of particles and, specifically, of the most abundant particles (protons, neutrons and electrons). Secondly, atoms and nuclei are usually satisfactorily described by the nonrelativistic approximation, that is, the well-developed methods of nonrelativistic quantum mechanics are applicable to them. Both these considerations bring atomic and nuclear physics close to macrophysics.

The natural shift of the arbitrary boundary between microphysics and macrophysics is clearly illustrated by the following example. Before the microscope was invented, everything that could not be seen with a naked eye could be termed microscopic. Later the term microscopic was only applied to things that could not be seen with a microscope, for instance, individual atoms. Now, that atomic-scale and, to a certain extent nuclear-scale phenomena are reasonably well understood, and may be readily visualized in the mind, there are reasons to reserve the term "microscopic" for those things which can be "seen" only poorly, or not at all [one can even observe individual atoms directly now with a field-emission ion microscope(50) or with a special electron microscope(51), and recent advances in this field(173) are even more impressive]. Thus, it is basically clear that microphysics must include the field which was, and still is called elementary particle physics, though now it is more often called high-energy physics or, more specifically, meson physics, neutrino physics and so on.

Hence, the subject matter of microphysics is mainly the "primary", "elementary" particles, the interactions between them, and the laws that govern them.

This definition of microphysics is not absolute and, to a certain extent, is even arbitary, as most definitions tend to be. But at least, it seems to be no less clear or admissible than other definitions. Anyway, the term "microphysics" is used here in this sense. This definition makes microphysics, as in the past, the field of research in which the very fundamentals are not clearly understood, let alone the secondary issues. As for the basic laws, microphysics (in the above sense) is at present mainly a domain of relativistic quantum theory. Finally, in terms of dimensions, a characteristic length in microphysics is of the order of, or smaller than, 10^{-11} cm (for electrons the Compton length $\hbar/mc = 3.85\times10^{-11}$ cm and for baryons $\hbar/Mc \sim 10^{-14}$ cm). (A classification based on the type or nature of laws applicable in the field, seems to be the most profound one. Thus, the most consistent classification in physics would be to distinguish between three fields: the first governed by classical laws, the second by non relativistic quantum mechanics, and the third by relativistic quantum theory. These fields may be called macrophysics, microphysics, and for example, ultramicrophysics, respectively. But the most consistent approach is not always the most convenient and widely accepted one. Therefore, it seems best to use the old terms of microphysics and macrophysics, but to shift the boundary somewhat between them.)

It should be noted, too, that macrophysics still does not include the whole of nuclear physics (in essence this has been discussed at the end of Sec. 9).

The study of nuclei makes a significant contribution to our knowledge of the interactions between nucleons, and of those between nucleons and other particles; the relativistic effects are fairly important here and, generally, there are close and numerous ties between nuclear physics, and the physics of elementary particles. Thus, I am, probably, somewhat ahead of time when I break with tradition, by counting nuclear physics as a part of macrophysics. But such a classification has hardly any real significance unless one believes that those who work in microphysics are the salt of the earth, and that macrophysics is a second rate science. Personally, I, of course, do not subscribe to such strange views although not uncommon, in some quarters; I believe (as many others do, of course) that a man should be judged only on his merits and not on the position he occupies. In ancient Russia noblemen were very keen to observe the rules of precedence at the tzar's court. But there is no tzar in science, and a scramble for special priorities seems to be out of place here (if the reader considers these remarks to be superfluous, I refer him to Dyson's paper[(2)], for example).

Work on the fundamental problems of microphysics involves difficulties which are similar to those encountered by the scientists who developed relativity theory and quantum mechanics. Studies of this calibre, even if they yield comparatively modest results, put a great stress on a scientist, demanding from him exceptional efforts and imagination; they give rise to a unique atmosphere charged with emotion; passions ride high... (This atmosphere can be best depicted in fiction, and I cannot give a really good example. I can cite Einstein's words which concluded his lecture on the history of the development of the general relativity theory, which dramatically illustrate the spirit of fundamental studies[(52)]: "In the light of knowledge attained, the happy achievement seems almost a matter of course, and any intelligent student can grasp it without too much trouble. Behind are the years of anxious searching in the dark, with their intense longing, their alterations of confidence and exhaustion and the final emergence into the light—only those who have experienced it can understand that".) But we have deviated from our subject; I want to state only that I cannot describe adequately the problems of microphysics in all their variety and depth. But this is not my aim; I describe below some of microphysical problems; their selection is arbitrary to an even greater extent than in other parts of this book, and their description is necessarily brief. It may be just this feeling of dissatisfaction with the microphysical part of the book which made me write this section, and Sec. 16, and that was, probably, not the wisest decision. Fortunately, competent discussions of microphysical problems are not rare, and I shall refer the reader to them below.

11. Mass spectrum. Quarks and gluons

Only three elementary particles were known before 1932, namely, the electron, proton and photon. Later studies led to the discovery of neutrons, positrons, muons, pions, heavier mesons, hyperons, resonance particles, electron neutrinos, muon neutrinos, and antineutrinos. Some of these particles are no less (though no more) elementary than the proton or electron. Others (for instance, hyperons and resonance particles) seem rather like excited states of lighter particles. The majority of particles are unstable; they transform into one another and are surrounded by "clouds" of virtual particles (for instance, nucleons are covered with pion "blankets"). Thus, the very concept of an elementary particle has become very complex. The particles are described by their mass, spin, electric charge, lifetime and some other properties and quantum numbers[53, 54] and the number of different particles is fairly large.

Some fifteen years ago the above statement seemed indisputable. Now it must be qualified to a significant extent. Of course, physics made a very important step in going over from the concept of several stable or long-lived particles (such as neutrons) to the picture of hundreds of particles (most of which are short-lived). But back in 1963–1964 there appeared the hypothesis of quarks, which are the prototype particles from which all baryons and mesons are built (the particles of both these types have strong interactions and are known as hadrons). The concept of quarks has become widely accepted in recent years, particularly, after the discovery in 1974 of new particles[55] whose properties could be successfully interpreted within the framework of the quark model, including the quarks of the fourth type known as charmed quarks (originally, quarks of only three types were postulated; see below). Thus, we can say that many years of studies of baryons and mesons, their nature and structure have resulted in the development of a new quark model of the structure of these particles.

When the quark hypothesis was put forward opinions on it diverged widely. Firstly, there were general doubts as to (see below) whether questions of the type "what is a proton made of?" are valid. Secondly, quarks are typically assumed to have fractional electric charges 2/3 and—1/3 (the unit charge is the charge of a positron or proton). But such fractional charges have never been observed and seem strange. Moreover, the intense search for quarks started after 1964, has not yielded results. Of course, it is extremely difficult to state absolutely that something does not exist. But it seems very likely (and most probable, at the moment) that quarks cannot exist in a free state as individual particles as can baryons, mesons or leptons. This would seem to give sufficient grounds for doubting the physical existence of quarks. But the quark model has not only not been abandoned, but is becoming increasingly accepted.

I cannot discuss the quark model in detail here. Readers who are interested in it may read the papers cited in ref. 56 (most of them are fairly popular). In our short discussion of quarks, we shall start with a table containing the quantum numbers of the quarks of four types known as flavours.

Flavour (quark type)	Charge	Baryon number	Strangeness (s)	Charm (c)
u (up)	2/3	1/3	0	0
d (down)	−1/3	1/3	0	0
s (strange)	−1/3	1/3	−1	0
c (charmed)	2/3	1/3	0	1

All quarks have a spin of 1/2 and, hence, they are fermions. A baryon consists of three quarks; for instance, protons and neutrons have compositions uud and udd, respectively. Strange and charmed quarks s and c are contained only in strange and charmed particles. For antiquarks all quantum numbers change their signs; for instance, the antiquarks $\bar{u}$ has charge —2/3 and baryon number —1/3. Mesons consist of a quark and an antiquark. For instance, the positive pion has the composition (configuration) $u\bar{d}$ (obviously, the charge of such a configuration is $2/3 + 1/3 = 1$, the baryon number is $1/3 - 1/3 = 0$, and the spin can be zero as it should be). Unfortunately (?), the above four particles and four antiparticles have proved to be not enough. A quantum number known as colour had to be introduced, so that a quark of each flavour can be in states described by three different colours (arbitrarily, red, green and blue). Three quarks comprising a baryon must be of three different colours so that the baryon has no colour. Mesons also have no colour since the colour of an antiquark corresponds to the anticolour of the quark.

Thus, the total number of quarks and antiquarks is 24, if we take into account colour. But this is not all. Even now, experimental results suggest the existence of quarks of a fifth flavour, and quarks of a sixth flavour have been introduced in theory. For quarks of the fifth and sixth flavours the baryon number is 1/3 and the spin is 1/2, as for quarks of the other four flavours (see table on p. 40). The fifth quark b (known as the bottom or beauty quark) has a charge of —1/3, and a mass of the order of 5 GeV (the mass of the c quark is of the order of 1.5 GeV and is, apparently, much higher than the masses of the u, d and s quarks). (Since quarks do not exist, or in any case, have not been found (with any degree of certainty), in a free state the concept of mass for them is somewhat abstract or extrapolative. The measured mass of hadrons is the sum of the masses of the quarks (and antiquarks) comprising them, and the mass corresponding to the binding energy.) As mentioned above, there is experimental evidence for the existence of the b quark of the fifth flavour. On the contrary, the t quark of the sixth flavour, with a charge of 2/3 (known as the top or truth quark), has not yet been found and there is not even indirect evidence for its existence (as, indeed, can be said about quarks bound in hadrons). The reason for this seems to be that the mass of the t quark is greater than 15 GeV, that is, much greater than even the mass of the b quark and therefore, hadrons containing t or $\bar{t}$ quarks cannot be produced in existing accelerators[174] (however See p, 46)

If there are six flavours and three colours of quarks then the total number of quarks and antiquarks is as large as 36. Suggestions have been made to increase the number of flavours and colours of quarks. At any rate, one cannot state that the quark model is necessarily limited to 24 or 36 quarks and antiquarks. Suffice it to say, that quarks interact with each other and this interaction involves exchanging quanta of some fields (as electromagnetic interaction involves exchanging photons). The quark model needs not one, but several, (typically, eight), fields mediating the interactions between quarks. The quanta of these fields are known as gluons (from glue). Thus, the total number of particles in the quark model of matter comes up to several dozens.

Is it not too many? This question tends to crop up when advantages of the quark model are discussed. By itself such an objection is hardly significant; even if the number of quarks and gluons is large, reduction of hundreds of hadrons to combinations of quarks, even of several types, brings some order and elegance into the picture.

A much more significant and fundamental question is whether we can meaningfully discuss particles (quarks) that are not observed in a free state? What is the meaning of the statement that a baryon "consists" of three quarks? There is, however, a clear answer to the last question—a proton scatters electrons and neutrinos as if it contains three point particles (consists of three particles) known as partons, which could be quarks[57].

But this does not prove the existence of quarks. For instance, a magnetic needle, or any magnet, behaves as if there were magnetic poles at its ends. In reality, no magnetic poles exist (at least under normal conditions) and magnetism is reduced to currents (motion of electric charges) and dipole (spin) magnetic moments of some particles (electrons, protons, etc.). This analogy between magnetic poles and quarks goes a long way[58]; any division of magnets leaves the poles "paired" (that is, any magnet, however small, has two poles), and similarly, no known transformation of hadrons gives rise to individual quarks, so that quarks can be produced only in the form of baryons and mesons, that is, in two's or three's. It should be noted also, that the quark model itself is still ambiguous. For instance, in some schemes quarks have integral charges, although substantial recent evidence is in favour of the fractional charge of quarks[175]. (It also has not been proved that mesons consist of a quark and an antiquark, rather than, say, two quarks and two antiquarks[150].)

The existence of quarks can be regarded as an aspect of the general question of whether we can distinguish between simple (elementary) and compound (complex) particles. We can state, for instance, that a hydrogen atom contains a proton and an electron since we can very easily break down (ionize) this atom, expending an energy of over 13.6 eV, very small compared with the 1 MeV needed for the generation of an electron-positron pair. Therefore, the number of particles in atomic physics is in practice conserved and a hydrogen atom can be broken down precisely to a proton and an electron which are stable particles in a free state.

Let us now consider the neutron. Does it consist of a proton and an electron as had been assumed long before neutrons were discovered, when they were regarded as the hypothetical "microatoms" of hydrogen? The answer is now known to be negative; the decay of a neutron is interpreted as the generation of an electron and an antineutrino and the conversion of the neutron into a proton ($n \rightarrow p+e^-+\nu_e+0.8$ MeV). We cannot say that a neutron "consists" of a proton, an electron and an antineutrino since a proton itself can decay into a neutron, a positron and a neutrino (although energy is consumed in this process, it can take place for protons in β^+ active nuclei). These illustrations demonstrate a limited application of the concept of "composition" to particles with large binding energies, or with decay products with high energy. But in general, this is just the case for the quark models of hadrons.

Comparatively high binding energies and, above all, the absence of free quarks (this is known as "confinement" of quarks) clearly hint that quarks are just auxiliary images (of the magnetic pole type in electrodynamics) that have no particular fundamental meaning, although they are convenient for describing various phenomena and properties of hadrons. This was just the opinion put forward by Heisenberg, one of the founders of quantum theory, at the end of his more than 50 years' work in physics(60). Physicists who are active in the field are also cautious in discussing the "existence" of quarks and their fundamental meaning(56b, 58).

But is it really necessary to think in terms of the alternative: "hadrons consist of quarks" or "quarks are just an auxiliary concept"? Would it not be more logical to describe hadrons as complex dynamic systems which have common features with atoms and atomic nuclei, but are qualitatively different from them, precisely because hadrons are indivisible into independent components? This seems to be a deep, and promising, concept. (I cannot miss the opportunity here to cite the definition of a deep statement attributed to Niels Bohr: "In order to define a deep statement it is first necessary to define a clear statement. A clear statement is one to which the contrary statement is either true or false. A deep statement is one to which the contrary is another deep statement(61).)

The development of atomic theory has invariably resulted in the introduction of new primary "building blocks" of matter (molecules, atoms, atomic nuclei, electrons, nucleons); the introduction of quarks would be yet another stage in this process, and then we would have to look for the components of quarks. Protoquarks (prequarks or preons) have already appeared in the physical literature under various names(176). But not everybody can believe in the "infinite matryoshka" (wooden, successively smaller hollow dolls, one inside the other)—open one doll, there is another one in it, open this one, there is yet another, and so on *ad infinitum*.

A concept which seems profound and, at the same time, quite natural according to some criteria, states precisely that infinite and mechanical repetition of the division of matter is stopped at some stage in a non-trivial way—baryons and mesons may simultaneously consist of some components (of the type of quarks) and not consist of them. This is possibly just a way of describing a situation in which the components cannot exist by themselves (in a free state) but nevertheless behave in some respects as nuclei in atoms or nucleons in a nucleus.

Similar considerations may perhaps provide some basis for reconciliation between proponents of quarks and those critics who only recognize as "really" existing those particles which can be free. But our ultimate concern is not with words or the modification and refinement of concepts and terms. The quark models currently serve as a springboard for the rapid development of the theory of quarks and gluons and their interactions, which attempts a quantitative description of the mass spectra of hadrons and their interaction (this theory is usually referred to as quantum chromodynamics).

If these attempts have far-reaching success, that is, a definite quark model of hadrons, then the controversy about the existence of quarks will somehow become meaningless. In other words, one will then be able to say that quarks are quite real although possessing properties that distinguish them from particles with which physics dealt earlier (primarily, the property of "confinement").

At present, the quark hypothesis can hardly be said to have fully and unconditionally triumphed. Moreover, doubts about this hypothesis are not totally groundless, as follows from the above discussion and, for example, from attempts to introduce prequarks (success of such attempts would mean that quarks only play some intermediate and perhaps auxiliary role). I should note, therefore, that other approaches to the problem of mass spectra have been, and are being, explored. One of them is the unified field theory of elementary particles[62], which can also be called a theory of "primary" matter, as it attempts to use as a basis a primary spinor field with a spin of 1/2. Ideologically, this theory is largely related to quantum chromodynamics and it has made a positive contribution[63].

But the development of a theory of primary matter derived from nonlinear equations for the field with a spin of 1/2 seems to be very attractive and an interesting attempt, even without any link with quantum chromodynamics and quarks. Another approach that should be mentioned is the attempt to develop relativistic models of particles with internal degrees of freedom[64].

Another unconventional and far from trivial approach to the problem of primary matter is the concept of Friedman's relating macrocosms (and even the Universe) to elementary particles. Following from the general relativity theory, the closed Friedman's world which may serve as a model for the Universe (see Section 19) has zero total mass and zero total charge (its mass is zero because the rest mass of the particles comprising the world is compensated for by the gravitational mass defect due to the gravitational interaction between the particles). If we consider a "semiclosed" world consisting, roughly speaking, of an almost closed world connected by a "neck" to an infinite space free of matter, the mass and charge of this world may be very small. (A clear, more detailed, explanation would be out of place in this book. The brief discussion here is meant only for those readers who are acquainted with the subject; those who are not may read the book[3b].) Thus, an entire world may look like a small particle, a friedmon, to a faraway outside observer. The mass, charge and other parameters (spin) of a friedmon are not defined in terms of classical theory and a quantum theory for friedmons has not yet been developed. It can be expected, however, that within the framework of quantum theory, friedmons will possess parameters of fundamental particles.

Of course, this is just a fantastic possibility so far; a hope, a dream. But the concept deserves careful analysis, if only for its daring in showing the way for uniting opposites, that is, the macrocosm and the world of fundamental particles. There are other, more specific, reasons for paying attention to this concept, particularly in the wider context of the relationship of microphysics to the general theory of relativity, and its quantum generalization which has not yet been developed (see also Section 19).

We have discussed above strongly interacting particles (hadrons, that is, baryons and mesons). The mass spectrum of leptons (electrons, positrons, muons, and neutrinos, as well as the recently discovered heavier tau leptons, with a mass about 1780 MeV[65],) is a separate, but no less important and fascinating problem. For instance, the reason for the difference in mass between the electron and the muon is still a mystery. We should also mention here the possible existence of some "exotic" particles that do not belong to the known particle families. Among such particles are magnetic poles (monopoles) and tachyons (hypothetical particles which travel faster than light in vacuum). Attempts to find monopoles and tachyons have failed. Such particles (particularly tachyons) most probably cannot exist, but it is typical of the current unsatisfactory state of the theory of fundamental particles that almost nothing can be forbidden.

Other hypothetical particles are maximons[3] and other particles[177] which have only gravitational interaction. We should specially mention the hypothesis of intermediate bosons, or W particles, responsible for the weak interaction. (Until fairly recently it was regarded as more probable that leptons interacted with hadrons "via weak forces", directly; but if appropriate intermediate bosons do exist then the weak interaction occurs in two stages—hadrons (baryons and mesons) interact with intermediate bosons and the latter interact with leptons.) If intermediate bosons exist, it is, to a certain extent, possible to unify the electromagnetic and weak interactions; that is, to reduce them to a single "primary interaction" (as a result the constants of the electromagnetic and weak interactions appear to be interrelated). At present such a unification based largely on the hypothesis of the existence of W bosons (these are three particles: $W^{\pm}$ and $W^0 \equiv Z^0$) is regarded as fully successful. Suffice it to say that in 1979 the Nobel Prize for physics was awarded for studies in the theory of unified weak and electromagnetic interactions[176, 178]. I value these results highly and think that the general basis of the theory is highly significant and correct. But it should still be remembered that $W^{\pm,0}$ have not yet been found experimentally (though, for an understandable reason; see Section 14) and the concrete unification theory[176, 178] has not yet been fully verified in some other respects (see Section 14).

As mentioned above, none of the attempts to solve the problem of mass spectra has met with unqualified success. It is true that the quark model has been increasingly accepted in recent years and, apparently, for hadrons, things have got going at last. But it is clear that a long and difficult path remains before the problem of the mass spectrum of particles is solved. This very extensive fundamental problem is one of the central problems of physics. (Of course, it deals not only with masses of particles but with all their properties. Moreover, the differences between baryons and leptons are qualitative and more significant than those between baryons with different masses.)

* (Note added in June 1983) In proton-antiproton ($p\bar{p}$) colliding beams with an energy of 270 GeV in each beam, the first results have now been obtained about the production (and, hence, about the existence) of $W^{\pm}$ bosons.[212] Experimentally the mass of these particles is $m_W = (81 \pm 5)$ GeV/c^2, whereas according to theoretical estimates $m_W = (82 \pm 2.4)$ GeV/c^2. The neutral Z^0 boson (which we called the W^0 boson above; this is, however, no longer done as it can lead to misunderstandings) which according to the theory has an approximately 10 GeV/c^2 larger mass, has also now been found[217]. Indirect indications about the existence of t quarks have also been published[213].

12. Fundamental length (quantized space, etc.)

The special and general theories of relativity, non relativistic quantum mechanics and the current theory of quantum fields make use of the concept of continuous, essentially classical, space and time (a point of space-time is described by the four co-ordinates $x_i = x, y, z, ct$, which may vary continuously). But is this concept always valid? How do we know that time and space "on a small scale" do not become quite different, fragmentized, discrete, quantized? This question is not at all new; the first to ask it apparently was Riemann[66] back in 1854 and it has been discussed repeatedly since that time. For instance, Einstein[67] in his well-known lecture "Geometry and experience" in 1921, said, "It is true that the proposed physical interpretation of geometry cannot be applied immediately to spaces of submolecular orders of magnitude. But nevertheless, even in questions of the constitution of elementary particles, it retains part of its significance. For even when it is a question of describing the electrical elementary particles constituting matter, attempts may still be made to ascribe physical meaning to those field concepts which have been physically defined for the purpose of describing the geometrical behaviour of bodies which are large compared to the molecule. Success alone can decide the justification of such an attempt, which postulates physical reality for the fundamental principles of Riemann's geometry outside the domain of their physical definitions. It might possibly turn out that this extrapolation has no better warranty than the extrapolation of the concept of temperature to parts of a body of molecular dimensions".

The question of the limits of applicability of Riemann's geometry, so lucidly expressed above (that is, in fact, applicability of macroscopic, or classical, geometric concepts), has as yet no answer. As we move towards the range of increasingly high energies and, hence, closer collisions between various particles (see Section 13), the scale of the unexplored space regions becomes smaller. We may now, apparently, state that down to distances(179) of the order of 10^{-16} cm the conventional space relationships remain valid or, more exactly, their application does not present contradictions. In principle, it cannot be ruled out that there is no limit at all, but it is much more likely that a fundamental (elementary) length $l_f \lesssim 10^{-16}$–10^{-17} cm exists, which restricts the applicability of the classical spatial description. Moreover, it now seems reasonable to assume that the fundamental length l_f is, at any rate, not smaller than the gravitational length $l_g = \sqrt{G\hbar/c^3} \sim$ $\sim 10^{-33}$ cm (see Section 19).

The problem of fundamental length has been discussed for many years in various forms and modifications (it enters into the theory(62) of primary matter, various versions of the theory(68) of quantized space, and so on). This problem is closely linked to the problem of possible violations of causality in the processes involving elementary particles (violations of microcausality, as they are sometimes called), to some other problems of microphysics, and to the problem of singularities in the general relativity theory and cosmology (see Section 19).

If a fundamental length does exist it may, naturally, be assumed to be of significance in the solution of the problem of mass spectra. A fundamental length would probably serve as a "cut-off" factor which is, to a varying degree, required by the current quantum field theory; a theory including a fundamental length would, in principle, automatically get rid of divergent expressions. It is true that at present many divergent expressions can be successfully "regularized" (eliminated) and some people think that no fundamental length is needed at all for this purpose. Of course, this is, in a certain sense right, but is by no means evidence against the possible existence of fundamental length (see also Section 14).

The experimental search for fundamental length involves the study of collisions of particles at ever increasing energies, and "ultraprecise" measurements of various properties at lower energies. Generally, disagreement between experimental results and the predictions of a theory (such as quantum electrodynamics) would indicate a possible violation of the applied space–time concepts and the need for the introduction of fundamental length, and we shall touch on this subject in Sections 19 and 21.

13. Interaction of particles at high and superhigh energies

The interactions of particles at high and superhigh energies are studied for a variety of purposes: to probe the structure of particles and of space itself on a small scale, to find new resonances (excited states of baryons and mesons), and to determine the energy dependences of cross-sections for elastic and inelastic scattering.

In nucleon–nucleon collisions the smallest separation is

$$l = \frac{\hbar}{m_\pi c} \frac{Mc^2}{E_c} \tag{10}$$

where $\hbar/m_\pi c \sim 10^{-13}$ cm is the Compton length for pions, M is the nucleon mass ($Mc^2 \approx 1$ GeV), and E_c is the nucleon energy in the centre-of-mass system of reference. For more detail see, for instance, Ref. 69. If one nucleon is at rest and the other has an energy $E = \sqrt{M^2c^4 + c^2p^2}$ then

$$E_c = \sqrt{(E+Mc^2)Mc^2/2} \tag{11}$$

The highest energy obtained in accelerators is $E \approx 500$ GeV (Batavia, USA) and even in this case we have $E_c \approx 15$ GeV and $l \approx 5\times10^{-15}$ cm. Particles with energies up to $E \approx 10^{20}$ eV can be found in cosmic rays, but in practice it will be scarcely possible to use cosmic protons with energies exceeding 10^{15} eV for the analysis of individual collision events; for these protons we have $E_c \lesssim 10^3$ GeV and $l \gtrsim 10^{-16}$ cm (however, some data(180) on the interaction of particles in cosmic rays can be obtained for higher energies, particularly for $E \lesssim 10^{18}$ eV). The colliding beam technique makes use of the head-on collisions of particles. If each particle has an energy E' and rest mass M then the centre-of-mass system coincides with the laboratory system and $E_c = E'$. Colliding proton beams with $E' \approx 30$ GeV have been already produced in Europe. See also p. 46), In terms of the same values of E_c this experiment is equivalent to using a proton beam with energy

$$E = \frac{2(E')^2}{Mc^2} - Mc^2 \approx 2000 \text{ GeV} \tag{12}$$

hitting a stationary hydrogen target. Evidently, colliding beams with particle energy E' in each beam can give rise to states (particles) with energies of up to $2E'$. In the foreseeable future we may expect to obtain energies $E' = E_c \sim 1000$ GeV corresponding to an energy

$$E \approx \frac{2(E')^2}{Mc^2} \sim 2\times10^6 \text{ GeV} = 2\times10^{15} \text{ eV} \tag{13}$$

But even then the separation, l^- is

$$l = \frac{\hbar}{m_\pi c} \frac{Mc^2}{E_c} \sim 10^{-16} \text{ cm} \tag{14}$$

so that distances less than 10^{-16} cm are very difficult to study even with the colliding-beam technique.

It should be noted in passing that the enormous difficulties involved in the study of particles with high energies, things such as the very short lifetime, stimulate the development of new methods and techniques for accelerating and detecting particles. Progress in the development of accelerators and detection devices (bubble and spark chambers, various counters, etc.) has been extremely impressive. In this connection it should be emphasized that, apart from its other influences, microphysics contributes to the expansion and diversification of experimental techniques and hardware in physics as a whole.

In collisions of non-strongly-interacting particles (muons, electrons, photons), the smallest separation is of the order of the wavelength in the centre-of-mass system, that is, $l = \hbar c/E_c = (\hbar/m_i c)(m_i c^2/E_c)$ where m_i are the masses of the colliding particles (here $E_c \gg m_i c^2$), so that the conditions for studying small distances are somewhat more favourable than for the case of collisions of nucleons. Moreover, very precise measurements, and careful comparisons with theoretical predictions, make it generally possible to "probe" distances smaller than the rough estimate above[179]. Nevertheless, it is perfectly clear how difficult it is to go beyond the threshold

$$l \sim 10^{-16} - 10^{-17} \text{ cm} \quad (\text{for } E_c = 1000 \text{ GeV the length } l = \hbar c/E_c \approx 1.5\times10^{-17} \text{ cm}).$$

As mentioned above, the main task of high-energy physics lies in the comparison of predictions of scattering theory with experimental results for ever-increasing energies, the study of new resonance states for baryons and mesons, and the determination of various effective cross sections[70].

High-energy processes involve primarily multiparticle production and not only the scattering and production of individual particles. Statistical, hydrodynamical and other methods[71] have been used in attempts to analyse the special features of multiparticle production. The above remarks mainly concern strongly interacting particles, hadrons (baryons and mesons), and therefore we should mention specially the interaction of muons and high-energy neutrinos with matter, particularly those produced in the Earth's atmosphere by cosmic rays (these are mainly neutrinos produced by the decay of muons and pions generated by cosmic rays[72]).

In contrast to the problem of mass spectra, and the problem of fundamental length, the interactions of particles at high and superhigh energies may seem to be only of secondary importance and to be lacking a definite and appealing physical goal. But it is not so, and if a reader really gets such an impression it is due only to imperfect discussion of the subject in this book. It should be mentioned that all the problems of microphysics mentioned above are, in fact, closely interrelated and highly interdependent. In discussing interactions of high-energy particles separately, I would like once more to emphasize that the physics of high-energy particles is by no means confined to the problems of mass spectra and fundamental length. For instance, the energy dependence of various interaction cross-sections for different particles (especially at superhigh energies or, formally, for $E \to \infty$ and, primarily for strongly interacting particles) are of very profound and, to a certain extent, independent, significance for theory. Moreover, studies of interactions between particles at increasingly high energies are very important for the verification of hypotheses[73] about the existence of a variety of particles, among the most significant of which are, perhaps, the $W^{\pm}$ and $Z^0 \equiv W^0$ vector bosons responsible for weak interactions (See p. 46)

14. Violation of CP invariance. Unified theory of weak and electromagnetic interactions. Grand unification theory. Proton decay. Neutrino mass.

In 1956 it was found that parity was not conserved in weak interactions. Specifically, in the beta decay of radioactive nuclei whose magnetic moments are aligned by an external magnetic field, different numbers of beta particles are emitted along the field and perpendicular to it (^{60}Co nuclei were used in the first experiments. This means that decay proceeds differently in the right-handed and left-handed systems of co-ordinates and hence does not exhibit parity invariance. But in all decay processes observed up to 1964, *CP* invariance was observed, that is, all interactions were invariant under the simultaneous operations of charge conjugation C (replacement of a particle with antiparticle) and space inversion P. Violation of parity may be explained as follows: a particle (a neutron, for example) possesses an "intrinsic screw" so that its decay is different in the direction along the "screw" from that in the opposite direction. Then CP invariance means that if a particle has a "left-handed screw" its antiparticle has a "right-handed screw".

The discoveries of violations of C and P invariance, and confirmation of CP invariance, were of the utmost importance and increased still further the interest in weak interactions. What is more, the weak interactions did not "disappoint their fans"; a discovery was made in 1964 whose significance appears to be very great, although far from being quite clear as yet. This was the decay $K_2^0 \rightarrow \pi^+ + \pi^-$ ($K_2^0 \equiv K_L$ is a long-lived neutral K meson which decays in this case into negative and positive pions) which only occurs with violation of CP invariance[74, 210]. Thus, CP invariance can be violated in nature, although it should be noted that all known processes that violate CP invariance are three orders of magnitude less probable than CP invariant processes. Violation of CP invariance seems to suggest a result of fundamental importance, namely that forward and reversed times are not equivalent. The fact is that CPT invariance follows from the first principles of the current theory for all interactions. This means that interactions—and, of course, their effects—remain unchanged under a combination (in any order) of the following three operations: space inversion P, charge conjugation C (replacement of a particle with its respective antiparticle), and time reversal T (replacing t with $-t$). The fundamental significance of CPT invariance is quite clear from the fact that it provides an identity for the masses and lifetimes in decays for particles and antiparticles. There is no evidence of violations of these properties of particles and antiparticles.

If an interaction exhibits CPT invariance but not CP invariance, it means that it is not T invariant, that is, it is not invariant under time reversal. But both classical theory (mechanics, electrodynamics and general relativity) and quantum theory (mechanics and field theory) satisfy the requirement of time reversibility; formally, this means that their equations are invariant for the operation of replacing t with $-t$, so that, with the appropriate change in initial conditions, a process proceeds along the same pathway and via the same states as before, but in the reversed direction.

The irreversibility (non-equivalence of the future and the past, or non-invariance under replacement of t with $-t$) observed in macroscopic physics, chemistry and nature is due to the complexity of macroscopic objects (great numbers of particles are involved) and thus final states tend to be less ordered than initial ones. So, under the conditions of *CPT* invariance (which is the most probable case, although it has yet to be rigorously proved) the discovery of violation of *CP* invariance meant the discovery of *T* non-invariance in fundamental interactions and processes. We may interpret this result as we did for violations of *C* and *P* invariance. Indeed, if we regard elementary particles as very complex entities (in a sense, this is precisely so; see above) we can imagine that a particle possesses not only an "intrinsic screw" but also an "intrinsic clock" running in a definite direction.

Studies of violations of *CP* invariance in weak interactions are continuing and it would be especially interesting to go over to higher energies. It is possible, for instance, that at high energies and correspondingly (to a certain extent) small space–time intervals violations of *CP* and *T* invariance are no longer "small" (say, their probabilities are comparable to the probability of weak interactions).

The above discussion follows, on the whole, my discussion of this subject in the 1974 edition of this book. At that time I thought that violation of *CP* invariance occupied a special place among the problems of weak interactions. It should be emphasized that this problem remains unsolved and significant (here, as almost always, I omit the obvious qualifications: "in my opinion" or "as far as I can see"). But currently the focus of interest in the field of weak interactions is, of course, the unified theory of weak and electromagnetic interactions. (This does not mean that the violation of *CP* invariance is a forgotten subject; the close interest in it is illustrated by the award of the Nobel Prize in Physics in 1980 precisely for its discovery[2, 10].)

Back in the 1930s it was suggested that weak interactions were due to the exchange of intermediate vector $W^{\pm}$ bosons, just as electromagnetic interactions could be regarded as being due to the exchange of photons. In this respect a deep analogy can be found between electromagnetic and weak interactions. But two very significant difficulties were encountered here. The photon mass is zero and photons are well known. The mass of the intermediate W bosons must be very large and they have not yet been found (it is reasoned that since their mass is so large they, of course, cannot be produced in the accelerators available).* Thus, the concept of intermediate W bosons was regarded as just another of many hypotheses lacking proper substantiation. But in 1967 a theory appeared which treats photons and W bosons with a unified approach and explains the difference in their masses[175, 176, 178].

*W bosons were first observed at Cern Geneva in 1982—1983 (See p. 46).

I have successfully resisted the temptation to attempt a superficial presentation of the elegant concepts forming the basis of the unified theory of weak and electromagnetic interactions (generalized gauge invariance, spontaneous symmetry breakdown). I refer readers to the popular papers[75] and a paper[63] which is intelligible not only to theorists, describing the relationship between gauge theories and superconductivity, and which is very helpful for understanding the theory (see also Ref. 181). Reviews[77, 177] can be useful reading on gauge theories. These are developing rapidly and are now applied not only to weak and electromagnetic interactions, but also to strong and gravitational interactions[76].

The following two aspects should be noted in this connection. Firstly, the advantages of the unified theory[75, 176–178] of weak and electromagnetic interactions became obvious only a few years after its formulation (these consist, primarily, in elimination of divergences, or renormalizability of the theory). Secondly, a significant aspect of the theory is the introduction of an intermediate vector neutral $W^0 \equiv Z^0$ boson in addition to charged $W^\pm$ bosons. The exchange of such a neutral particle, even in the first approximation, results in scattering processes which are not predicted by the same approximation from the theory which considers only $W^\pm$ bosons (this is the case, for example, for the scattering of the muon neutrino ν_μ by an electron, e, and for the scattering of the muon neutrino and the "normal" electron neutrino ν_e by protons or neutrons). In the parlance of theorists, processes involving Z^0 bosons are known as processes with neutral currents. The experimental evidence for the existence of neutral currents was first obtained in 1973 and more evidence has been obtained in later years[75,176–178] This was an indisputable triumph for theory. Now some other predictions of the theory are being discussed and verified[78]. But the theory can hardly be regarded as fully substantiated until $W^{\pm,0}$ bosons are found. According to some estimates, the mass of $W^\pm$ boson is between 58 and 68 GeV[79]. Masses of 79.5 GeV ($W^\pm$) and 90 GeV (W^0) with a level width of 2.6 GeV have also been predicted[182]. The masses of $W^{\pm,0}$ bosons may prove to be different, but there is no reason to think that they will be of a different order of magnitude; thus, these particles can be produced in the next generation of accelerators[80]. (In order to produce a particle with mass m, the energy of accelerated particles must be greater than mc^2 in the centre-of-mass system. Therefore, even if the energy of accelerated protons colliding with a resting proton is $E = 500$ GeV, the particles produced in the accelerator can have an energy of only $mc^2 \leqslant E_c \approx 15$ GeV; see Section 13.)

Apart from $W^{\pm,0}$ bosons, gauge theories (particularly those attempting simultaneous treatment of weak, electromagnetic and strong interactions) introduce other particles, in particular, scalar particles. Unfortunately, the mass of some of these may prove to be enormous (up to 10^{16} GeV or more[79, 176–178]), so that verification of their existence may be a matter of several decades. But this would scarcely affect the fate of gauge theories as a whole since any theory leaves some problems to be solved. It is more important to note that gauge theories which include strong interactions are still far from complete and are still evolving. The gauge theory, which only considers strong interactions, that is quantum chromodynamics, has made very significant progress. It can generally explain and describe the behaviour of quarks at a short distance from each other (for example, in protons) where they can be regarded as almost free particles which cannot move apart to large distances (these properties are often known as asymptotic freedom and confinement, or infrared slavery, of quarks). Another significant result from existing theory describes the effect of macroscopic factors, primarily very high temperatures on elementary particles. and vacuum[63]. Various gauge theories, particularly those including quarks with new flavours, yield significant results for weak interactions (and violation of *CP* invariance[79, 81]). At present, gauge theories seem to be the main trend in the development of microphysics, but it would be naive to think that nothing unexpected and no new difficulties will be met along the way.

Now I should like to note the following. Before the unified theory of weak and electromagnetic interactions was developed there were singularities in the theory of weak interactions (that is, it was not renormalizable) and some relationships had had to be "cut off" at an upper limit of energy, E, or equivalent length $l = hc/E$. Under such conditions there was some inducement to suggest the existence of a fundamental length $l_f \sim 10^{-17}$ cm. There is now no reason to introduce such a length. But does this mean that the final estimate of the fundamental length is now given by the gravitational length $l_q = \sqrt{G\hbar/c^3} \sim 10^{-33}$ cm (see Section 19)? In my opinion, it would be too hasty to come to such a conclusion. After all, the difference in the two estimates above of fundamental length is 16 orders of magnitude, and though recent advances in the theory of weak interactions can be regarded as very important, they present no evidence against the existence of a fundamental length $l_f \gg l_g \sim 10^{-33}$ cm (see Section 19).

Though it was something like 20 years ago that quark models, quantum chromodynamics, and unified theories of weak and electromagnetic (as well as strong and gravitational) interactions started to develop, it is only in most recent years that they have become especially wide-ranging, yielded important results, and given rise to great expectations. In general, after a period of relative depression, microphysics now experiences a period of elation governed by enthusiasm and inspiration. It is only the future that can show objectively the real significance, and meaning, of the current stage of development of microphysics, and its results, although some of the people involved in building the new microphysics have apparently valid reasons for making reliable predictions even now.

I do not work actively in microphysics now and, of course, I cannot say that my opinion is sufficiently well substantiated or significant. Nevertheless, I must note that after I had read many papers in the field (some of which are cited here) in order to prepare this edition, my feelings were quite different from those I had when reading about many sensations and fashionable fields of research in the past. I think that quarks, gluons, gauge fields, etc. are not just a current fad, a new bright hope of physics, or some limited successful result, but provide something fundamental, if not final, to our understanding of nature.

In the last Years, my opinions as expressed above, have strength ened. New developments in microphysics are clearly a breakthrough (the current period of brilliant development may be compared to the periods between 1924—1925 and 1930—1932 when quantum mechanics was created and strengthened). I have noted repeatedly above that I cannot give proper attention to the problems of microphysics in this book. Nevertheless, I have made for this English translation some additions to the list of references and to the text, in order, at least partially, to discuss these problems. Here is the largest of these additions.

The successful advance of the unified theory of weak and electromagnetic fields[(75, 176–178)] is indisputable but, as noted above, for reliable verification of the theory $W^{\pm,0}$ bosons must be discovered. The theory now encounters another difficult question. According to the unified theory, the relationship between weak and electromagnetic forces should result in some small, but qualitatively new, effects in atomic physics. Specifically, the parity of interaction between electrons and nucleons must not be conserved. This must result in the rotation of the plane of polarization of light passing, for instance, through bismuth vapour in the frequency range of some atomic transitions (if parity is conserved this rotation is exactly zero). Appropriate experiments were carried out in Oxford (Britain), Seattle (USA), and Novosibirsk and Moscow (USSR). At present, the British and American results seem somewhat inconclusive, the Novosibirsk results agree fully with theory[183], and the Moscow results sharply contradict theoretical predictions[184]. According to the Novosibirsk results[183], a parameter R describing the angle of rotation of the plane of polarization is $-(20.2\pm2.7)\times10^{-8}$, while the theoretical value of R is about -18×10^{-8}. According to Moscow results[184], the plane of polarization practically does not rotate: $R=-(2.3\pm1.3)\times10^{-8}$. What is to be done? The answer is clear: new experiments should be carried out in other laboratories. Apparently, this will be done soon.

* According to the most recent data known to the author a rotation of the polarization plane of light realy occurs on passing through bismuth vapour. In that case $R = -8$ to 9×10^{-8} which apparently is not in contradiction to theory.

Rotation of the plane of polarization of light in the vapours of heavy elements is a separate problem; it is, so to speak, off the main direction of the assault launched on a unified theory of all interactions. A basic component of this theory is quantum chromodynamics, which deals with quarks and gluons. But while quarks are already "familiar" particles, this cannot be said of gluons, neutral vector particles, the exchange of which mediates the interaction between quarks. Gluons, as quarks, most probably cannot exist in a free state (although there is practically no doubt about this, it is extremely difficult to be absolutely sure in such cases). Significantly, in 1979 fairly reliable although indirect experimental evidence was obtained for the existence of gluons in studies of particle generation in colliding e^+e^- beams[185].

Advances in the unified theory of weak and electromagnetic interactions, on the one hand, and the results of the theory of strong interactions (quantum chromodynamics) on the other, contribute to the development of a unified theory of these three interactions (only the gravitational interaction is not included). This is known as grand unification. It is typically based on three types of quarks (quark doublets (u, d), (c, s) and (t, b), where each quark can have one of three colours) and three types of leptons (doublets (ν_e, e), (ν_μ, μ) and (ν_τ, τ)) with their respective antiparticles (all particles have spin 1/2, that is, they are fermions). Among these 24 particles (we do not include the antiparticles) only t quarks have not been found, although evidence for the existence of the neutrino ν_τ, corresponding to the τ lepton is rather indirect[65]. Grand unification theory groups all these particles together with a number of scalar (spin 0) and vector (spin 1) bosons, taking into account some conditions of symmetry and gauge invariance, but it is still far from being complete and definite[176–178, 186, 187].

There is much that is still unclear to me here, and I shall therefore not go into details. But we should note some of the main qualitative results from grand unification that seem to follow naturally from quite general considerations. Indeed, if quarks and leptons are somehow grouped together they can generally convert into one another, and make a contribution to the mass of all particles.

This leads to a fascinating possibility that the proton is (or, more exactly, can be) unstable! Indeed, proton decay of the type $p \to \pi^0 + e^+$, for instance, is quite feasible from energy considerations. If the baryon number is conserved, then such a decay is forbidden, but the possibility of the conversion of quarks into leptons and *vice versa* corresponds to the nonconservation of baryon number. Available experimental data indicate that the mean proton lifetime T_p is more than 10^{30} years (recall that the "age of the Universe, that is the time of its observed expansion, is only of the order of 10^{10} years; see Section 19). There are approximately $N = 10^{34}$ nucleons in 10^4 tonnes ($= 10^{10}$ g) of water and, if the probability of decay of a fixed neutron is approximately the same as that of a proton, for $T_p = 10^{31}$ years the number of decays observed in a year in this amount of water should be $N/T_p = 10^3$. However, grand unification theory has still not predicted an accurate value for T_p. In some modifications of the theory $T_p \to \infty$ (the proton is stable), but in another modification T_p is of the order of 10^{31}–10^{33} years.

Experiments are now being prepared to measure T_p, and in the largest of these installations 10^4 tonnes of water are used (this is why we made the calculations above).Thus, if T_p is of the order of 10^{31} years it can be measured, but if T_p is more than 10^{33} years it will probably take many more years to find an answer to this question.

If proton decay is observed it will be a triumph for the grand unification theory, but negative results will by no means lead to its refutation, as can be seen from the discussion above. If $T_p \lesssim 10^{33}$ years then strong, weak and electromagnetic interactions become equivalent at an enormous energy, $E_x \sim 10^{15}$–10^{16} GeV, corresponding to a mass $m_x = E_x/c^2 \sim 10^{-9}$–$10^{-8}$ g (the proton mass $m_p = 1.6 \times 10^{-24}$ g).

It is such a high value of E_x that provides for a low probability of proton decay. Note that the so-called gravitational or Planck mass (the mass of maximon) $m_g = \sqrt{\hbar c/G} = 2.2 \times 10^{-5}$ g ($E_g = m_g c^2 \sim 10^{19}$ GeV) is just 3–4 orders of magnitude larger than the above mass m_x. The mass m_g corresponds to the length $l_g = \hbar/m_g c = \sqrt{G\hbar/c^3} = 1.6 \times 10^{-33}$ cm, while the length $l_x = \hbar/m_x c \sim 10^{-29}$–$10^{-30}$ cm. Hence, grand unification assumes the absence of any fundamental length $l_f \lesssim 10^{-29}$ cm (see Section 12).

Grand unification (note once more that it is far from being complete) should be followed by a unification theory for all interactions, including gravitational interactions. Within the framework of certain theoretical concepts (in the absence of a fundamental length larger than l_g) this will involve working with lengths in the range $l \sim l_g \sim 10^{-33}$ cm, masses in the range $m \sim m_g \sim 10^{-5}$ g and energies in the range $E \sim E_g \sim 10^{19}$ GeV $=$ $= 10^{28}$ eV (in the above discussions we sometimes quote masses in electron-volts (eV) which is, of course, quite admissible; here, for the sake of clarity, we distinguish between mass m and rest energy $E = mc^2$).

Intense efforts are underway to unify all the interactions, and thus to fulfil Einstein's dream of a genuinely unified field theory. We cannot here go into details of the cosmological or other implications[177, 194] (see, however, below, and Section 19).

The relationship between neutrinos and other particles which reflects their "unification" generally results in neutrinos having a non-zero rest mass m_ν (this mass can, of course, differ for ν_e, ν_μ, and ν_τ neutrinos). In its current state the theory cannot predict this rest mass, and even if it could the neutrino mass would still have to be measured by experiment. This is by no means a new suggestion. There were originally two reasons for assuming the neutrino mass to be zero (here we mean the electron neutrino ν_e). Firstly, experiments have shown that this mass is small in the sense that $m_{\nu_e} \ll m_e = 5.1 \times 10^5$ eV. Secondly, the assumption that the neutrino mass is zero makes the theory simpler and more elegant than for the case for which it is non-zero. Of course, the inadequacy of such arguments was obvious, and experiments were undertaken which yielded the limit $m_{\nu_e} < 50$ eV $\sim 10^{-4} m_e$. Such experiments are based on studying the spectrum of beta decay. Tritium decay is convenient for this purpose ($t \rightarrow {}^3He + e^- + \bar{\nu}_e$) since the spectrum boundary for it is very low ($E^{e, max} = 18.6$ keV). Recent, allegedly more accurate, results [188] have yielded a range of between 14 and 46 eV for the mass of the electron neutrino. Clearly, new experiments, carried out by different groups, are needed. Another indication[189] that neutrinos have non-zero rest mass was obtained, in particular, from analysis of the reactions

$$\bar{\nu}_e + d \begin{cases} \nearrow n + n + e^+ \\ \searrow n + p + \bar{\nu}_e \end{cases}$$

In fact, they involve the so-called neutrino oscillations, that is, the conversion of an electron neutrino ν_e into neutrinos of other types (ν_μ and ν_τ) and *vice versa*. If such oscillations occur then the masses of mutually converting neutrinos are different and hence at least one of the masses is non-zero. In experiment, these oscillations must lead to a variation of the intensity of a beam of, for example, electron neutrinos *in vacuo* with distance, even if it does not diverge. This effect is very significant for the interpretation of the results of detection of solar neutrinos (see Section 23) and, of course, for fundamental theory. If the neutrino mass $m_\nu \gtrsim 10$ eV, this is of enormous significance for cosmology[190] (see Section 19). But if the masses of neutrinos of all types are smaller than 1 eV their contribution to cosmology is generally insignificant. For physics it is, of course, essential to know the masses of neutrinos of all types irrespective of their actual values. Determination of the neutrino mass is undoubtedly one of the most important and urgent problems in microphysics.

*According to the most recent data known to the author the proton lifetime (referring to the $p \rightarrow \pi^0 + e^+$ decay channel) $\tau_p > 10^{32}$ years (somewhat earlier a value $\tau_p > 6.5 + 10^{31}$ years has been discussed[214]). So far no reliable data have been obtained indicating the occurrence of neutrino oscillations or a non-zero neutrino mass. Neither are there any reliable data indicating the existence of massive magnetic monopoles (this problem has been discussed extensively in theoretical papers).

15. Non-linear phenomena *in vacuo* and ultrahigh electromagnetic fields

The problems of microphysics discussed above are very wide in scope; each of them could be subdivided into a number of more concrete partial problems. But it is not my aim to increase the number of problems or to make their classification elegant or consistent; this would run contrary to my more modest goal which is to review briefly the general state of affairs in physics. Thus, we shall discuss the processes in ultrahigh electromagnetic fields as an example of a more special, but highly important, problem.

The peculiar behaviour of matter in ultrahigh magnetic fields has already been discussed in Section 6. In contrast to magnetic fields, a strong electric field generally disintegrates atoms. For instance, if the strength ϵ of the external electric field is of the order of the field of the nucleus (proton) of the hydrogen atom at a distance of the atomic radius $a_0 = h^2/me^2 = 5\times 10^{-9}$ cm, that is, if

$$\epsilon \sim \frac{e}{a_0^2} \sim \frac{e^5 m^2}{\hbar^4} \sim 10^7 \quad \text{esu/cm} \approx 3\times 10^9 \text{ V/cm} \tag{15}$$

then the atom disintegrates very rapidly. Actually, such disintegration (atom ionization in an external electric field) also occurs in weaker fields, but in fields considerably lower than 10^8–10^9 V/cm the lifetime of a hydrogen atom is long enough. A uranium atom will be completely and rapidly stripped of its electrons only if the external field is as high as $10^9 Z^3 \sim 10^{15}$ V/cm (the charge of the uranium nucleus, eZ, is $92e$, and the radius of the uranium K shell is $a_{0,Z} \sim \hbar^2/me^2 Z \sim 10^{-10}$ cm, so that the field of the nucleus at this shell is of the order of eZ^3/a_0^2).

In still higher electric fields it is not only the electrons of the heaviest atoms that cannot "hold out", and are stripped from their nuclei and accelerated by the field, but also the vacuum itself. The fact is that a real (physical) vacuum is not at all "empty"; a vacuum is polarized by the field and can generate various particle pairs. The easiest to produce are the lightest pairs, that is, electron–positron pairs. A field ϵ_0 whose energy at a distance of the Compton wavelength $\hbar/mc \sim 3\times 10^{-11}$ cm is of the order of the rest energy ($2mc^2 \sim 10^6$ eV $\sim 10^{-6}$ erg) of such pairs, generates them at a high rate. Hence we have $e\epsilon_0\hbar/mc \sim mc^2$ or

$$\epsilon_0 \sim \frac{m^2c^3}{e\hbar} \sim 10^{14}\ \text{esu/cm} \approx 3\times10^{16}\ \text{V/cm} \tag{16}$$

Weaker fields can also produce pairs at a fairly high, though not as catastrophic, rate. Therefore, the generation of pairs in a vacuum would probably be observable in fields of strength of about 10^{14} V/cm. A variety of other interesting effects also occur in such high (and sometimes in weaker) fields. It should be noted that particles with a high energy E can generate pairs at a high rate in an electromagnetic field which is lower than the field (16) by a factor of E/mc^2 (in the frame of reference linked to the particle, the field is E/mc^2 stronger than in the laboratory frame of reference[82]).

The highest fields produced so far in a laser beam focus are of the order of 10^9–10^{10} V/cm (see Section 7). To produce a field of 10^{14} V/cm the laser power and the electromagnetic energy flux at the laser beam focus must be raised by ten orders of magnitude, a task that is apparently unrealistic for physics today. But for electrons with energy $E \sim 20$ GeV (that has already been reached) we have $E/mc^2 \sim 4\times10^4$, and thus such electrons will produce photons and pairs in a laser beam field of about 10^9 V/cm in the same way as from electrons initially at rest in a field of about 4×10^{13} V/cm.

Ultrahigh electric fields occur in the vicinity of atomic nuclei. For instance, the field at the boundary of the uranium nucleus is of the order of $eZ/R^2 \sim \sim 3\times10^{16}$ esu/cm $\approx 10^{19}$ V/cm ($Z = 92$, and the radius of the nucleus is $R \sim 10^{-12}$ cm). But this field at the uranium nucleus cannot generate pairs; higher fields are needed for that and quantitative analysis[83] indicates that pairs can be generated by fields of nuclei with $Z > Z_c \approx 170$. Such nuclei can be produced only for a short time in collisions of two nuclei for which $Z_1+Z_2 > Z_c$. But nevertheless, this process can be quite interesting and, of course, the problem of vacuum polarization and pair generation near superheavy nuclei has other important implication[49, 63, 83, 84]. Finally, it is especially interesting to analyse the generation of pairs near singularities in cosmological solutions describing the evolution of the Universe (see Section 19).

Vacuum is polarized not only by a high electric field but also by a high magnetic field; the corresponding characteristic field value is $H_0 \sim \dfrac{m^2c^3}{e\hbar} \sim 10^{14}$ Oe (that is, the same as for an electric field; see eq. (16)). In a magnetic field with H comparable to H_0, and especially in a field for which $H > \mathrm{H}_0$, vacuum behaves as a non-linear anisotropic medium and strongly affects the propagation of electromagnetic waves (that is, in quantum language, the motion and general behaviour of photons; see, for instance, Refs. 1b, 82 and the references cited therein). The effect of a high magnetic field on a vacuum ceased to be an abstract problem after it had been found that near the surface of pulsars magnetic fields could be as high as 10^{13} Oe (see Section 21).

16. Microphysics yesterday, today and tomorrow

All things are in a state of flux and changing, and the changes occur not only in the subject matter of the science we refer to as microphysics, but also in its status among other sciences and specifically in physics. One has only to look through physical, abstracting and popular science journals to arrive at the conclusion that the proportion of microphysical problems in all these publications has shrunk considerably over the past 20–25 years. Unfortunately, no accurate quantitative data are available, but in my opinion* the ratio of the numbers of papers on microphysics to those on macrophysics is currently lower than 25 years ago, by at least an order of magnitude. Other indices of scientific activity (the number of graduate students specializing in the field, the number of conferences held, and so on) would probably present a similar picture.

I believe that the primary reason for this is that in the recent past—from 30 to 35 years ago—microphysics occupied an exceptional place among other sciences; and now things have changed.

Microphysics deals with the most fundamental, essential and, therefore for many, most attractive problems in physics. There has been no change in microphysics in this respect. But up to the middle of this century microphysics exerted a decisive influence on the development of the natural sciences in general. Indeed, at the time microphysics was mainly concerned with the study of atoms, and later with atomic nuclei. The development of many fields of physics, astronomy, chemistry and biology depended on the powerful impetus which was given by understanding the structure of atoms, and the laws governing their behaviour (to understand them quantum mechanics had to be developed!). In a similar way, studies of

*In this connection one cannot help deploring the fact that in the USSR too little attention is paid to statistical (or any other) analysis of the development of trends in science. It should also be noted that this decrease in the share of microphysics cannot be ascribed to the fact that I have classified the main part of atomic and nuclear physics as macrophysics. Suffice it to recall that such branches of microphysics as high-energy physics, meson physics, neutrino physics, etc. were non-existent in the past. Yet my definition fully retains the vanguard position of microphysics in physics as a whole (see Section 10).

atomic nuclei resulted in the use of nuclear (atomic) energy, providing a reason for calling the twentieth century an atomic age. (It is other question that this aspect of nuclear physics was not appreciated for some time.)

In the great majority of cases, the physicists working on microphysical problems were not concerned with the practical applications of their results; the source of their persisting enthusiasm was interest in the problems themselves, the urge to know "how the world runs" and the unquenchable desire to overcome difficulties and to reach the truth. But a high concentration of effort, the wide scope of the work, and the support and attention provided by society—in particular, by the scientific community—were due largely to an awareness of the significance of microphysics for the development of natural sciences as a whole, and for humanity in general, as a means of solving outstanding practical problems.

Now the situation is quite different. Microphysics deals with particles that either live for minute fractions of a second or, as with the neutrino, pass almost freely through the Earth and can only be detected with enormous difficulty.

Of course, the scientific significance of a problem cannot be gauged either by the lifetime of particles, or by their penetration capacity. Current problems in microphysics are no less mysterious and difficult than the problems of the past. In other words, microphysics still is (and under the above definition will always be) the most advanced and fundamental area of physics; its vanguard, let's say. It should be noted that this opinion, which I share, is not indisputable (see, for instance, Ref. 4). Many problems of macrophysics or, for example, biology, are very deep and independent; their solution is not made easier by the fact that the relevant fundamental laws (for instance, non-relativistic quantum mechanics) are already known. Yet the difference between microphysics and macrophysics seems to be significant enough for microphysics and, say, cosmology to be singled out (see Section 19). But of course this does not mean that other natural sciences are looked upon as something second-rate or non-fundamental.

But what has changed is the character and significance of the subject matter of microphysics. In the past, microphysics was concerned with "everyday things", that is, atoms and atomic nuclei; now it studies outlandish and rare animals (at least, by terrestrial standards).

[There are, of course, some exceptions. For instance, muons live for microseconds and are of some interest for chemistry and, maybe, even for building fusion reactors making use of the so-called muon catalysis; specifically, muons in deuterium or in a deuterium–tritium mixture facilitate reactions[1]. Moreover, studies of protons and electrons are, of course, continuing but they are too detailed (for instance, the quark model of the proton) to be directly relevant to our understanding of the problems of atomic and nuclear physics.] But, as mentioned above, the literally domineering position of microphysics in science was, to no small extent, due to the exceptionally high priority of the problems it dealt with.

Thus, in my opinion, the role played by microphysics both in physics and in all natural sciences in general has changed radically, and I believe that this change will persist if not for ever then, at any rate, for a very long time (this assumption is the most controversial one).

Dispensing with scientific language, I would say that in the first half of this century microphysics was the first lady of natural sciences. Now and in the future microphysics is and will remain "merely" the most beautiful lady. But this is just the point: opinions about the most beautiful lady may differ, while, by definition, there can be only one first lady (for instance, the President's wife). In my eyes microphysics was and still is the most beautiful lady in physics. But in contrast to some of my colleagues I believe that adoration should not be accompanied by a neglect of changes in age and character, and ignoring other objects worthy of admiration.

These remarks may seem quite commonplace ... but only to those who agree with them. It is precisely because they are controversial that they are made here. I became aware of this some 20 years ago when I wrote something along these lines in a different context(85). Some of the objections and critical remarks aimed at me at that time were, though, a result of misunderstanding or egocentricity. For instance, some people understood the words about the changing and, to a certain extent, decreasing role played by microphysics, if not as a call to stop the construction of high-energy accelerators and general support for microphysical research, then, at least, as a justification of such measures. It goes without saying that I meant nothing of the kind. [I cannot help adding here that I detest attempts to link closely (or even to tie up) the discussion of the development and planning of science to special narrow interests, and to problems of a given research field under specific local conditions. The problems of financing, construction, etc. in the development of science depend on many factors, among which purely scientific considerations may sometimes be of secondary significance by comparison to, for example, economic or technological ones. There are even fewer reasons for drawing practical conclusions from only scientific considerations, without analyzing comprehensively the scientific organizational problem under discussion. The situation would change considerably if the funds available for the development of science were increased manyfold by, for instance, diverting the enormous sums of money wasted by mankind on various unproductive activities such as smoking and drinking. But something of this kind can happen only in a science-fiction novel.]

There is, however, one essential objection which is worth discussing. In the early stages of nuclear research development, the prospects of nuclear power production were far from clear and their evaluation was sometimes quite wrong. Such examples are not rare. Generally, it is hard and sometimes even impossible to make predictions about the development of science. Therefore, it seems possible or even fairly probable that, if one takes into consideration a number of analogies, microphysics will assume once again its role as a generator of large-scale problems on a par with the problem of nuclear energy. (An example of this is the hope of introducing quark catalysis[151] for which free quarks are needed). Then, naturally, the relative significance of microphysics could dramatically increase.

Of course, such a possibility cannot be completely ruled out. The fact that there is a chance, albeit remote, of new, important and useful discoveries must in itself be a sufficient reason for encouraging the development of microphysics in all possible ways, in addition to the interests of "pure" science.

At the same time, even if we admit that a reversal can occur in the practical importance of microphysics in the future, the above remarks on its present-day position are by no means contradicted. Moreover, I fail to understand why it is regarded by some as heresy or bad manners to suggest (I do it without hesitation) that the most glorious time of microphysics has perhaps ended (in terms of its effect on the development of society, technology, etc.).

Unfortunately, as regards the prospects of microphysics, I have almost no hope of proving my case. On the other hand, I shall hardly have a chance to confess my errors, for even optimists tend to recognize that no radical change in the role played by microphysics in science and technology can be expected within the lifetime of the present generation.

Incidentally, the prestige of microphysics is still extremely high, and only those who were spoilt by getting high-priority attention may feel any dissatisfaction with the situation since they have had to move away from the limelight. It is only in this respect that the status of microphysics in physics has been somewhat altered, as it has been "shouldered aside" by astrophysics (including space research) and, especially, by biology. Surely, the dramatic progress in biology that we are currently experiencing (more exactly, the progress in some fields of biology, such as molecular biology, biophysics, etc.) has not only a great scientific significance, but also opens up fantastic prospects for solving such major human problems as the elimination of diseases (in particular, cancer), considerable prolongation of the natural human lifespan, artificial "test-tube" life, the tapping of unused brain resources, and so on. On the other hand, astrophysics is a science that is deeply fascinating in itself irrespective of the prospects of its useful applications which are generally quite remote and uncertain. In this respect, the position of microphysics at present and in the nearest future is largely similar to that of astrophysics. Clearly, the construction of large accelerators is no less essential than the construction of large telescopes on Earth and on satellites.

Some of the above remarks and suggestions are very similar to those made by Dyson(2) (by the way, Dyson's paper and the first version of this book appeared almost simultaneously but are, of course, completely independent). I would like to cite here three rules suggested by Dyson for resolving critical situations encountered in physical research work (applicable on the scale of a laboratory or an institute):

Don't try to revive past glories.

Don't do things just because they are fashionable.

Don't be afraid of the scorn of theoreticians.

The first two rules are fairly obvious and, moreover, Dyson comments on them. But there might be some misunderstanding about the third rule, particularly if the reader does not know who the author is.

Physicists are customarily divided into theorists and experimentalists. Ideally, an experimentalist sits at the apparatus he designed himself and makes measurements. In addition, he must get money, materials and instruments for building his apparatus, supervise the work of technicians and assistants (there are sometimes many of them), and interpret experimental data. This goes on and on, sometimes for years, and the only result of all this work may be a more accurate value of a parameter or a constant (for instance, the magnetic moment of proton, the mass of a resonance particle, etc.).

As for the theorist, he, ideally, sits at his desk in a tidy, well-lit room overlooking a garden or a lake or, at worst, lounges on a sofa at home and meditates on the "nature of things" or performs some calculations making occasional breaks for exciting discussions of various scientific and general topics.

Both the above concepts are, of course, quite abstract, but they help to understand why there is often little love lost between the abstract experimentalist and the abstract theorist. In real life things are not as simple as that. As recently as the nineteenth century there was no clear-cut distinction between experimentalists and theorists. Naturally, some physicists performed more experiments, while others made more calculations, depending on their tastes and skills, but most of them were apt to do both. It was the increasing sophistication of experimental techniques, the dramatic rise in the number of physicists, the growing competition between them, the increase in work rate and in the rate of information exchange, that gave rise to a sharp division of labour in physics and, to a certain extent, carved separate niches for theorists and experimentalists. The results of this differentiation of roles are ambiguous. The statement that physicists have ceased to understand each other no longer sounds paradoxical or absurd; sadly it is too often true.

But why talk about physicists in general when even among theoretical physicists one can meet all grades of specialization, starting with mathematicians who somehow got interested in solving physical problems, to those down-to-earth physicists who for some reason do not work with their hands, or who lost their experimental connection by chance. Naturally, theorists belonging to opposite poles of their own group, let alone theorists and pure experimentalists, quite often fail to find a common language and distrust each other.

Now, if one reads Dyson's third rule without a prior knowledge of the author's personality, one might visualize an exasperated experimentalist: theorists got on his nerves by lecturing him on what to do and how to work, and by hinting at his ignorance of "true" physics. In fact, Dyson is one of the well-known contemporary theoretical physicists. It is only his knowledge of the manners exhibited by a fraction of his fellow theorists that prompted his advice not to be afraid of their scorn. This is a manifestation of his fondness for his "trade", rather than a betrayal of it.

Genuine theoretical physics is an integral part of physics as a whole; it cannot even exist without experimental physics, let alone dominate it. A theoretical physicist is not a prophet or priest; more often than not he is just a lucky chap free of those troubles which incessantly pester an experimental physicist. This is why scorn or ridicule by theorists can be only counterproductive (the same, of course, can be said about disrespect and distrust towards theorists shown, if not in words then in deeds, in some experimental quarters).

Of course, we are talking about exceptional cases but they justify exercising the right of self-defence, the more so since I heard that "Dyson is a defector" and "Ginzburg is an enemy of nuclear physics", and all these charges were caused just by the above remarks! I would not pay attention to them, or mention them, if only my hurt feelings were concerned, nor would I attempt to answer the criticism. My reasons for discussing it here are quite different; my aims are to stimulate discussion, perhaps by making it more heated, to induce readers to think, to work out their own opinions and to express and defend them fearlessly, but especially not to be indifferent and unconcerned. However, indifference is, perhaps, better than intolerance and disrespect of unacceptable opinions of "outsiders", and egocentric protection of one's own views and interests by defaming one's opponents. At the same time, there is nothing that can benefit the development of science more than enthusiastic, friendly discussion, debate and argument and, fortunately, these prevail in science.

This section is somewhat polemical in tone and therefore I have decided not to make any major changes in it for this edition, but to assess again my opinion after several years. It is a natural thing to do in view of the brilliant advances in microphysics made in recent years (substantiation of the quark model by the discoveries of new, in particularly charmed, particles; the development of the unified theory of weak and electromagnetic interactions, etc.). Moreover, despite my (quite sincere) protestations of love for microphysics, some (though few) readers still suspect that I underestimate it. I would like such readers to read this section once more.

Great new advances in microphysics have not changed my opinion at all. They could have changed it since this opinion concerns not current microphysics itself, but its relation to other fields of physics and other sciences which have remained unchanged. Maybe I should only repeat that I think that the position of microphysics at present is similar to the position of astrophysics. And there is no better position!

I should add that it is, of course, very good when science is useful for industry, agriculture, communications, medicine, etc. but demands that science should produce immediate practical benefit seem to be unfair and unjustified. Firstly, very often practical use of scientific results cannot be directly evident, and are found only many years later. Secondly, for many people research work is personal fulfilment and a lifework just as music, art or poetry are for others. Then why should scientists be subjected to more stringent requirements as regards practical usefulness than musicians? Of course, the funding of a human activity strongly depends on the chances of practical return, but this is a quite different question. In general, I would like to emphasize that my opinion about the currently less significant role of microphysics and astrophysics for society (in comparison with, say, macrophysics or biology) should by no means be regarded as any kind of a reproach.

III.

Astrophysics

17. Experimental verification of the general theory of relativity

Einstein put forward the general theory of relativity (GTR) in its final form in 1915. By that time he had suggested his three famous ("critical") effects to be used for verification of the theory, namely, gravitational displacement of spectral lines, deflection of light rays in the gravitation field of the Sun, and displacement of the perihelion of Mercury. More than half a century has passed, but the problem of the experimental verification of GTR is still as urgent as ever.

Why is this so?

All the effects predicted by Einstein have been observed, but the experimental accuracy is still low[86, 87]. For instance, the error in measurement of the gravitational displacement of spectral lines is about 1% and, moreover, the effect itself does not depend on the type of gravitational theory. (Recently the agreement between theoretical prediction and the measured displacement of lines was improved by two orders of magnitude[152] but this success has no fundamental significance[87]).

The deflection of light rays in the gravitational field of the Sun was first observed more than 60 years ago (according to GTR the deflection is as high as 1.75″ if the light travels in the immediate vicinity of the solar disk). Unfortunately, available optical techniques make such measurements possible only during solar eclipses. Although this effect has been discovered, and agrees with theory, the measurement error is about 10%. Such poor experimental accuracy until recently provided a justification for putting forward alternative gravitational theories which differ from Einstein's theory. But the deflection of radio waves, instead of light, can be measured to verify the theory. The deflection of radio waves from quasars passing near the Sun has been measured and found to agree with GTR to within 1%.

Similar and even better accuracy was obtained in the measurement of the relativistic delay of radar signals, reflected from Venus and Mercury and passing near the Sun. This relativistic delay time is about 2.10^{-4} s (this effect is physically equivalent to the deflection of light in the gravitational field of the Sun; the signal passed near the edge of the Sun and, of course, the reflecting planet was in the superior conjunction). The displacement of the perihelion of Mercury was measured to about 1% and agreement between these results and the GTR predicitions was for many years regarded as the best substantiation of GTR (apart from results on the identity of gravitational and inertial masses which are accurate to 10^{-12}). But it was suggested more than 15 years ago that agreement only seemed to be good because the quadrupole moment of the Sun was not taken into account. This objection, which seemed at first somewhat spurious, was given some support by observations which were interpreted as indicating flattening of the Sun. But the current view is that the Sun's flattening is so insignificant that the resulting quadrupole moment cannot have a noticeable effect on the motion of Mercury.

Thus, we can now only say that even for weak gravitational fields, that is, for small parameters $|\varphi|/c^2$ (on the surface of the Sun $|\varphi|/c^2 = GM_{sun}/r_{sun}c^2 = 2.12 \times 10^{-6}$) GTR has been verified only with an accuracy of 1%. This is poor accuracy for modern physics and provides, if not a reason, then an opening for the discussing alternative gravitational theories.

The lack of accurate experimental verification of GTR is explained by the small size of effects observable from the Earth and generally within the solar system, and by the relatively low accuracy of the astronomical methods used this purpose. But now new prospects have been opened up by the launching of interplanetary probes (space rockets), and the use of radio and other techniques, which predict verification of GTR to an accuracy of 0.1–0.01% or better(86). For instance, the Viking probe orbiting around Mars was used to measure the relativistic delay of signals(191) which agreed with the GTR prediction to within about 0.1%.

Provided that experimental measurements in the solar field support GTR (and I strongly hope they do), the problem of verification of GTR will move on to another, quite different stage. This will be the verification of GTR validity in strong fields, or near and inside supermassive space bodies, let alone in cosmology.

The last two phrases were written about ten years ago when the question of the Sun's flattening was still under debate, and the deflection of light and the delay of signals in the Sun's field were measured to several per cent. Now, the results for all three predicted effects in weak fields agree with the theory to within 1% and even to within 0.1%, thus making it of a higher priority to verify the validity of GTR in strong fields. I have presented my view of the subject in a paper(87a) (see also Sections 19–21 and 24).

But does this mean that no further verification of GTR in weak fields is needed? The answer is apparently no, since any evidence (which needs, of course, to be reliable and proven) of even the smallest deviations from GTR predictions within the solar system and in weak fields generally would be of exceptional significance. Most physicists (and I am amongst them) think that this is extremely unlikely. But it is really of no use to talk about the probability of discoveries in such situations. It would be more consistent to talk about the "mathematical expectation" of a discovery equal to the product of the probability of a discovery and its significance. The mathematical expectation of finding deviations from GTR would be considerable even if their probabilities were negligibly small. Anyway, such discussions do not lead anywhere.

Obviously, no advances can be made in GTR verification without making new observations and measurements. We are eagerly awaiting them, particularly those concerning the "black holes" (See Sections 21 and Ref. 87).

18. Gravitational waves

Any relativistic theory of gravitational fields must provide for the existence of gravitational waves *in vacuo* similar to electromagnetic waves. In the GTR this similarity goes even deeper since in this theory gravitational waves are purely transverse. The concept of gravitational waves *in vacuo* appeared simultaneously with the GTR, and the well-known and widely used equation for the intensity of gravitational radiation by masses whose velocities were much smaller than the velocity of light (see eq. (110.16) in Ref. 88) was derived by Einstein[89] back in 1918.

Gravitational waves must be emitted by any mass whose quadrupole mass moment is non-zero and varies with time (for fast-moving bodies, multipole moments of a higher order also make a contribution). Binary stars and planetary systems are the simplest space objects satisfying these conditions.

Gravitational interaction is the weakest of all known interactions. The macroscopic manifestations of gravitation that we know so well from our everyday life are significant because the masses involved (the mass of the Earth, for example) are very large. (For two protons, the gravitational attraction is weaker than the electrostatic repulsion by a factor of $e^2/GM^2 \sim 10^{36}$, where $G = 6.67 \times 10^{-8}$ cm^3 g dm^2 is the gravitational constant, $e = 4.8 \times 10^{-10}$ esu is the proton's charge, and $M = 1.67 \times 10^{-24}$ g is the proton's mass.) Therefore, it is hardly surprising that the intensity of gravitational radiation is typically (say, for binary stars) comparatively low, and its detection rather difficult.

No observation of gravitational waves has yet been made with any certainty, and the prospect of receiving gravitational waves from binary stars and pulsars seems to be very small. Suffice it to say that even if the PSR 0531 pulsar in the Crab nebula emitted gravitational waves with intensity $L_g \sim 10^{38}$ erg/s, then the gravitational radiation flux at the Earth would be only $F_g \sim 3\times10^{-7}$ erg/cm^2 s. The sensitivity of available gravitational wave detectors at present corresponds to minimum fluxes of 10^4–10^6 erg/cm^2 s, that is, lower by at least 11 orders of magnitude than is needed in the above example(86, 90, 91). (The value of L_g given above is the total luminosity of the Crab nebula in the entire spectrum of electromagnetic radiation. I think that there is no reason to expect that the pulsar gravitational radiation intensity is so high; it is probably lower by a few orders of magnitude. Possible exceptions may be X-ray pulsars and collapsars (see Sections 20 and 21) in close binary systems for which the quadrupole moment of a rapidly rotating compact star might be anomalously large.) In order to employ available techniques for detecting pulsar gravitational radiation with fluxes as high as 3×10^{-7} erg/cm^2 s a detector weighing several tons must be cooled down to 10^{-2}–10^{-3} K. (If the detector is made, not from aluminium, but from a material of the type of sapphire, with very low internal friction, its mass may be smaller by a few orders of magnitude.) It can be done but, of course, with great difficulty.

Fortunately, apart from periodic gravitational radiation from binary stars and pulsars, we can expect fairly powerful bursts of gravitational radiation (with a duration of the order of 10^{-4}–10^{-3} s) to be produced by the collapse of stars (for instance, resulting in a neutron star or a black hole, that is, in some supernova outbursts) and by the collisions of neutron stars and black holes(86, 90, 91). The energy W_g liberated in such events (in the form of gravitational radiation) is generally not greater than 10^{55} erg $\sim 10M_{sun}c^2$. But in our galaxy supernova outbursts occur, on the average, once every 10–30 years. We should therefore aim at detecting radiation from other galaxies. For instance, if we take a sphere with radius $R \sim 3$ Mpc $\sim 10^{25}$ cm containing approximately 300 galaxies, pulses of gravitational radiation with time-integrated flux $F_g \sim W_g/4\pi R^2 \sim 10^4$ erg/cm^2 will be received from it several times a year (on average, supernova outbursts in surrounding galaxies occur about once every 100 years). A more realistic estimate $F_g \sim 1$ erg/cm^2 corresponds to an energy $W_g \sim 10^{52}$ ergs and a distance $R \sim (3$–$5)\ 10^{25}$ cm.

It is extremely difficult, but still possible, to detect such bursts (pulses) of gravitational waves. Various techniques have been suggested which involve the use of satellites, lasers, superconducting magnetometers and other sophisticated apparatus of contemporary experimental physics(86a, 91). It is not clear how long we shall have to wait until cosmic gravitational radiation is detected, especially as estimates of radiation fluxes vary greatly (as we have seen above). Various laboratories work actively in this field and in the 1980 or '90s gravitational-wave astronomy can be expected to become a reality[192].

It is perhaps worthwhile to mention that the detection of cosmic gravitational radiation was reported some years ago[92]. Two massive aluminium cylinders (1.5 tons each), a distance of 1000 km from each other, were reported to vibrate at a frequency about 10^3 Hz under the effect of gravitational radiation coming from the direction of the centre of our galaxy. In the previous edition of this book I discussed these experiments and questioned the reliability of the results reported. It now is quite clear that no gravitational radiation with intensity corresponding to these results comes to the Earth from space. I recall this erroneous report not to criticize the author once more, or to boast of my critical foresight. It is generally easier to pass judgement than to do original work. As for the mistaken authors, they feel bad enough and it would be senseless to recall their errors constantly. On the contrary, I write here about the failed attempt[92] to detect gravitational radiation in order to emphasize the stimulating effect these experiments had. They generated great interest and their failure helped to push the problem of the detection of gravitational waves into the limelight. As at the end of Section 9, I am not speaking here against the careful checking and critical analysis which are, of course, necessary for science. Both extremes are harmful—excessive liberalism, and the not infrequent wholesale rejection of any unclear or insufficiently substantiated results. As almost always, the proper attitude is somewhere between these two extremes.

Some people link the detection of gravitational waves to the verification of GTR. This is correct, in a sense, particularly if one bears in mind that non-Einsteinian theories of gravitational fields predict not only transverse gravitational waves but also longitudinal ones, and in general, a picture different from that predicted by the GTR. It was reported[153] in 1979 that the variation of the orbit of the binary radiopulsar PSR 1913+16 agreed with the assumption that this system radiated gravitational waves in accordance with the GTR. My firm belief is, however, that the main aspect of the problem of detection of gravitational waves is that they can (and will) provide a new channel for obtaining important data on space. Therefore, gravitational wave astronomy is bound to appear sooner or later. And the sooner the better.

19. The cosmological problem. Singularities in the general theory of relativity and cosmology

Cosmology is the study of space and time on a large scale. Thus, cosmology is inseparable from extragalactic astronomy and covers a very wide range of other subjects. But the key problem in cosmology lies in understanding the evolution of the Universe with time, its character, and in developing a cosmological model that reflects reality (I assume here that the reader is acquainted with the basic consepts of contemporary cosmology and the landmarks in its development; this is justified to a certain extent by the ready availability of numerous publications in the field[87b, 88, 93]).

The homogeneous and isotropic cosmological models represent the Universe, in accordance with observational data, as an expanding system. Such models were first analysed by Friedman in 1922 and 1924, and later discussed by Lemaitre and many other scientists. Einstein was the first to suggest a relativistic cosmological model which was isotropic and homogenous, in 1917[94]. But this model was static. It corresponds to a single solution from the family of otherwise non-stationary solutions found by Friedman who assumed Einstein's cosmological constant Λ to be non-zero. When this constant is zero all homogeneous and isotropic models are non-stationary. Note that the local homogeneity and isotropy of space—and of the respective model — are compatible with global models differing in topology. There may probably be many cosmological models which have not yet been studied practically, but which agree with the available observational data[192a], in addition to the Friedman models. Interestingly, it was as late as 1934 that Milne and McCrea showed this non-stationary state to be classical in character; this means that it can be derived from the Newtonian theory of gravity (the fact is, that if only "attractive" gravitational forces act in a system of bodies, the system cannot remain at rest, and in the absence of rotation it will either contract or expand, depending on the initial conditions).

The mean density, $\rho(t)$, of matter in the Universe is a primary cosmological parameter (more precisely, it should be the energy density divided by c^2, since a contribution may be made by gravitational waves, to which we can hardly apply the term matter). In the isotropic and homogeneous (Friedman) models with $\Lambda = 0$ the model is closed for $\rho > \rho_c = 3H^2/8\pi G$ (an expanding and then contracting three-dimensional sphere) and the model is open for $\rho < \rho_c$. Here G is the gravitational constant and Hubble's constant H is now estimated at 75 Kg/s Mpc $= 2.4 \times 10^{-18}$ s^{-1} which corresponds to an "age" of the universe $T \sim 1/H \sim 10^{10}$ years. For this value of H the critical density $\rho_c \approx 10^{-29}$ g/cm^3; in the past, the value of $\rho_c(t)$ was higher, since H decreases with time. Determination of the density $\rho(t)$ or, more specifically, the density $\rho = \rho_0$ in our time has proved to be a very difficult task. The mean density due to visible objects (galaxies, quasars) is lower than ρ_c by a factor of about 15. But maybe the value of ρ_0 is determined by invisible components, that is, hot intergalactic gas (this is mainly ionized hydrogen and therefore very hard to detect), black holes, gravitational waves or neutrinos. For instance, if the neutrino mass $m_\nu \gtrsim 10$ eV, then intergalactic neutrinos formed in the past, when the Universe was sufficiently hot, can at present provide for a density as high as ρ_c, or even higher (to make a more accurate estimate one has to know the number of stable neutrino species and their masses, and to take into consideration neutrino oscillation, if any). But in any case if $m_\nu \ll 1$ eV (for all quasistable neutrinos), the neutrino contribution to density is small. It is, of course, highly meaningful and interesting that in this connection the problem of neutrino mass acquires primary cosmological significance(177, 190, 194).

Whatever the nature of the expansion of the Universe, obviously, it could not in the past proceed infinitely (I mean here an expansion whose rate does not tend to zero when $t \to -\infty$). Indeed, in all homogeneous and isotropic models expansion started either after a compression stage or at a moment $t = 0$ when the matter density was infinite (a singularity). If the cosmological constant $\Lambda = 0$, all solutions exhibit a singularity (solutions with $\Lambda \neq 0$ which do not have singularities, disagree with observational data).

A singularity (infinite density) seems to be logically admissible but many (including myself) believe that the emergence of a singularity indicates the inapplicability or limited applicability of a theory, or at least that something is not quite right with it. For a time some people hoped that singularities would appear in the Friedman models because of their high symmetry, and will disappear in non-uniform and anisotropic cosmological models (just as the image at the focus of a high-symmetry lens is blurred with its deformation). But these hopes proved to be wrong(88, 98); highly generalized anisotropic and inhomogeneous GTR solutions corresponding to cosmological models also have singularities (approximation to this point generally occurs in a rather peculiar oscillating fashion).

Thus, within the framework of GTR it seems impossible to get rid of singularities in the treatment of cosmological expansion or gravitational collapse (see the following section), This is not quite applicable to systems with non-zero total electric or bosonic charge, that is, fields of vector bosons(95). But this is by no means decisive evidence for the existence of real singularities with density tending to infinity. One should bear in mind the fact that GTR is a classical theory. But there are no doubts that the true (complete and consistent) gravitational field theory must be a quantum theory. Typically, these quantum effects are extremely small in astrophysics, as in most problems of macrophysics, but it is precisely in the vicinity of a singularity that quantum effects are greatly enhanced. Assume, for instance, that a fundamental length l_f exists (see Section 12). Then it seems almost certain that the classical GTR ceases to function at distances of the order of, or less than l_f and, probably, for densities $\rho \gtrsim \rho_f \sim \hbar/cl_f^4$ (the quantum constant $\hbar$ (g cm²/s), the velocity of light c (cm/s) and the fundamental length l_f

(cm) can yield only this density ρ_f (g/cm³)). When $l_f \sim 10^{-17}$ cm we have $\rho_f \sim 10^{30}$ g/cm³. It may be assumed that, under such conditions, densities exceeding ρ_f cannot be realized and the singularity disappears. If there is no fundamental length l_f not related to gravitation, then there will, nevertheless, emerge a gravitational length l_g (possibly it will play the role of the fundamental length l_f). Indeed, the gravitational constant G (cm³/g s²), the velocity of light c and the quantum constant $\hbar$ can be combined to yield the length

$$l_g \sim \sqrt{G\hbar/c^3} \approx 1.6\times 10^{-33} \text{ cm} \tag{17}$$

This length corresponds to a time $t_g \sim l_g/c \approx 0.5\times 10^{-43}$ s and the density

$$\rho_g \sim c^5/\hbar G^2 \sim \hbar/cl_g^4 \approx 5\times 10^{93} \text{ g/cm}^3 \tag{18}$$

Various analyses and estimates[93b, 96–98] indicate that, even in the absence of any fundamental length $l_f > l_g$, if one takes into account quantum effects the density cannot be higher, by an order of magnitude, than 10^{94} g/cm³. Under these conditions, apart from the growth of various fluctuations, particle pairs should generally be produced at a very high rate in the vicinity of the singularity. This suggests that classical singular GTR solutions cannot be extrapolated to densities exceeding ρ_g and, in general, to the singularity itself. Of course, we still lack a consistent quantum gravitational theory, not to mention quantum cosmology. Therefore, the limits of the applicability of a classical description are not quite clear. But this does not mean that quantum cosmology is not needed. The task seems to be exceptionally difficult, but it is of fundamental importance and must be accomplished. This problem is closely related to quantum effects for mini-black holes (see Section 21) which have been treated in numerous papers in recent years (see Refs. 93b, 96–98, and references cited in Section 21).

The above discussion was based on GTR, but many attempts are being made to solve cosmological problems without using GTR or, more exactly, by going outside its scope[99]. I should mention here the concept of a universe symmetrical in charges[100] which possibly does not contradict GTR but has another basis. None of these "unconventional" approaches to the theory of gravitation and cosmology has yielded any successful results, but they should not be rejected without detailed analysis[87, 193].

In astronomy, the cosmological problem and the related problem of singularities in GTR hold a place similar to that of microphysics in physics (as regards their character and the type of problem). Moreover, microscopic problems seem to interlock with problems of astrophysics and cosmology (not to mention the hypothesis of friedmons[3b, 95]). Apparently, our understanding of these problems depends on the introduction of new concepts, and the quest for truth in this field will generate many mistakes in repeated attempts to find the right way.

*We have already noted the relationship between cosmology (primarily, the processes at the early stage of cosmological expansion) and microphysics, and this relationship has been attracting increasing attention in recent years. The reasons for this are quite clear. As mentioned at the end of Section 14, the grand unification theories are currently concerned with distances of the order of $l_x \sim 10^{-29}-10^{-30}$ cm and, according to eq. (18), densities $\rho_x \sim \hbar/cl_x^4 \sim 10^{78}-10^{82}$ g/cm^3. But such densities could be observed (indirectly, of course) only at the earliest stages of cosmological expansion (the length $l_x \sim 3\times 10^{-30}$ cm corresponds to a characteristic time $t_x \sim c/l_x \sim 10^{-40}$ s). Moreover, the instability (decay) of protons and some other effects in grand unification theories affect the cosmological picture. In other words, the early Universe may be regarded as a unique laboratory of high-energy physics. I shall cite here only a few papers(63, 177, 194–198) out of the many publications in the field which, though arbitrarily chosen, can provide a lead for interested readers.

Undoubtedly, the union between cosmology and particle physics will be strengthened and extended in the future, yielding benefits for both sides.

*The problem of the early Universe in all its astronomical and physical aspects can at the present time claim the top place, if one determines this "place" by the number of publications, conferences, proceedings, and so on. This refers, on the one hand, to a "test" of a number of proposed unified field theories (we are thinking here of the GUTs, the Grand Unified Theories, and of the SuperGUTs, unified theories including gravitation) and, in general, to the "use" of cosmology to analyse fundamental physical theories(215). On the other hand, for the first time in many years, starting in 1981, the cosmological models themselves were significantly developed—we have in mind the appearance of the inflationary Universe model. Unfortunately, it is impossible to dwell here in more detail upon these models and we restrict ourselves to the remark that in the inflationary models the expansion of the Universe, very close to the cosmological singularity, is essentially different to what happens in the usual models of the expanding Universe (Big Bang)(215, 216). True indeed, the singularity itself remains. The latter has, however, for a long time been "under suspicion". There have been very many attempts to somehow "get rid of" the singularity and now such attempts have increased. In my opinion, just in this direction—along the path of cosmological models without singularities (oscillating models or some other models without a "true" singularity)—will there be further developments in the first instance.

20. Does astronomy need "new physics"? Quasars and galactic nuclei. Formation of galaxies

Can one expect to find deviations from the classical GTR solutions which occur everywhere, or locally, in space, apart from the early—that is, close to the classical singularity—stages of evolution of the Universe? One can formulate a similar question in a wider context if one considers not just deviations from GTR but the more general possibility of deviations from known physical laws in astronomy.

In a sense, this seems to reduce to the eternal question that is on the mind of many astronomers: does astronomy boil down to "terrestrial physics", that is, physics that is valid in our laboratories? A similar question has been troubling biologists for many years: can all biological phenomena be reduced to physics, that is, to molecular concepts, or not? (The evolution of views on this question has led to an increasing expansion of the "range of applicability" of physics in biology. The development of Bohr's views on this problem is an instructive example; see Ref. 85 and the references cited within.) Of course, there is no *a priori* answer to such questions. Apparently, the most natural—and in fact, most widely used—approach is as follows: let us apply the physics that is available without any restrictions; if we encounter really insurmountable difficulties we shall be ready to introduce new concepts and reconstruct or extend physical theories. It is probable that most people will agree with such an approach, but this will not be a consensus opinion for one has to agree on the definition of an insurmountable difficulty first!

In this respect physicists working in astronomy are typically much more conservative than "pure" astronomers (I mean here good, healthy, conservatism). Some astronomers seem to possess some need to escape from physical shackles; they exhibit a real yearning for speculation that is not restricted by known physical laws. As an illustration let me quote the words of Jeans[101]: "Each failure to explain the spiral arms makes it more and more difficult to resist a suspicion that the spiral nebulae are the seat of types of forces entirely unknown to us, forces which may possibly express novel and unsuspected metric properties of space. The type of conjecture that presents itself, somewhat insistently, is that the centres of the nebulae are of the nature of "singular points" at which matter is poured into our Universe from some other, and entirely extraneous, spatial dimension, so that to a denizen of our Universe they appear as points at which matter is being continually created".

These concepts of Jeans are sometimes invoked as if they were almost a prophecy. But they date back to 1928 when not much was known about the structure of galaxies and the theory of their evolution was practically non-existent (incidentally, the origin of spiral arms is now rather well understood).

Our current knowledge of galaxies is incomparably greater; for instance, they have been found to possess nuclei which play an important part, and are sometimes active[(102)]. But can one make from the available data such far-going conclusions as those of Jeans[(101)] and Ambartsumyan[(103a)] about nuclei being sources of matter, or that the nuclei are "a new form of existence of matter unknown to physics"[(103b)].

Most astrophysicists do not agree with these conclusions and there still remains some hope of explaining all the phenomena observed in galaxies, their nuclei and quasars[(102, 104)] without introducing fundamentally new physical concepts. Galactic nuclei and quasars may well be, or may contain in their central part, supermassive plasma bodies ($M \lesssim 10^9 M_{sun}$, $r \lesssim 10^{17}$ cm) with rapid internal rotary motion and magnetic fields[(104)] (magnetoids or spinars). Another possible explanation within the framework of GTR is that massive black holes are at the centres of galactic nuclei and quasars[(104)].

A similar explanation may be found for the problem of the missing mass in clusters of galaxies. Such clusters must be stable only if their total energy, which is the sum of their kinetic energy and the potential energy of gravitational interaction, is negative (the energy of gravitational interaction is negative, as it is taken to tend to zero with increasing distance between the masses). However, some clusters are clearly stable, while their total energy is positive if one takes into account only the known masses[(105)] (that is, primarily, the masses of the stars comprising these galaxies).

The picture would be changed if one found in the clusters some unknown masses making a sufficiently large contribution to the gravitational interaction to stabilize the clusters. The probable explanation for the missing mass is low-luminosity stars in the clusters, and intergalactic gas[(105)]. Another possible explanation involves collapsed masses (see Section 21). It has also been suggested that clusters can be stabilized by trapped neutrinos if the rest mass of the neutrino is non-zero but very small[(106, 190)]. But it is very difficult to find in clusters low-luminosity stars, intergalactic gas, collapsed masses or excess neutrinos, and no definite results have yet been obtained. However, the majority of physicists and astronomers believe that the missing masses will be found and the problem will be resolved to general satisfaction. But until it has been done there remains some scope for contriving such hypotheses as, for instance, the suggestion that known physical laws are violated in clusters (this concerns the laws of classical mechanics and the Newtonian law of gravitation since GTR effects in clusters are very small). If new matter is produced somewhere in clusters (for instance, in the nuclei of the galaxies comprising them) then missing masses are not needed to explain the results of observations, for in this case clusters might only appear to be stable (new galaxies are produced continuously and other galaxies move away from clusters)[103]. Thus, the problem remains unsolved, the associated difficulties can hardly be ignored but, at the same time, one can not really agree that a "lack" of mass in clusters indicates a need to alter fundamental physical laws.

But the references above to the opinions of the "majority" cannot help bringing to mind the words of Galileo, that in science the opinion of a single person may mean more than that of a thousand. Therefore, I do not at all propose that the majority rule be regarded as sufficient grounds for the unlimited application of known physical laws; I just state the facts. And the facts are that even the astronomical community, let alone the physical community, has not at all been convinced that fundamentally new physical concepts are needed for the understanding of processes occurring in galactic nuclei, quasars and clusters of galaxies.

The above discussion generally follows the text of the previous edition of this book. Now, several years later, one can hardly hear the voices of radicals attempting to introduce new physical concepts to explain the phenomena in quasars, galactic nuclei and clusters of galaxies. Of course, this is due to a better understanding of the nature of quasars, galactic nuclei and the dynamies of clusters. However, as in a few other places, I have not changed the text much, so that a reader can get an insight into trends in astronomy in the recent years.

There are some other aspects to be discussed concerning the possible discovery of new fundamental physical laws from astronomical observations. Here I shall note only the following: nobody questions the importance of new physical concepts. They are obviously needed in microphysics, and probably in cosmology (see Section 19) as well as, in general, in the vicinity of singularities (that is, the singularities occurring in the solutions of classical GTR—Einstein's non-quantum gravitational field theory). But there are no reasons at all to assert that new fundamental physical concepts and laws must necessarily be introduced, or discovered, for those regions or objects where conditions (density, temperature, etc.) are within the ranges covered by conventional physics. On the other hand, one cannot rule out the possibility of finding some essentially new phenomena under such conditions as exist in large-scale systems such as quasars, galactic nuclei and clusters, which may be due, for instance, to the presence of enormous masses, cosmic-scale distances, contributions of processes of very low probability, and so on. In other words, primary importance should be attached to the specific conditions prevailing.

The above conslusion is, of course, self-evident. Its aim, however, is to emphasize that relativity and incompleteness of our knowledge do not indicate the inescapable need to introduce new concepts and laws, as some people may do even if there is no evidence for the inapplicability of conventional physics (for more details see Ref. 107).

Thus, in my opinion it is highly probable that no new physics is needed to explain the processes occurring in galactic nuclei, quasars and clusters of galaxies (in contrast to the early stages of evolution of the Universe, that is, in the vicinity of the classical cosmological singularity). But it is precisely in galactic nuclei, clusters and quasars that the search is going on for deviations from GTR, conservation of baryon charge, etc. Studies of these systems are of outstanding significance both for the above reason, and because of their exceptional importance in astronomy (and cosmology) as a whole. These studies aim at understanding not only the structure and dynamics of these objects but their origins as well. The problem of the origin of galaxies and their clusters, though, emerged long before the discovery of galactic nuclei and quasars. That this problem is to a certain extent independent is evidenced by the fact that not all galaxies have a pronounced nuclei. The causes of formation of galaxies and their clusters within the framework of the general expansion of the Universe, are still being discussed actively.[102a, b; 105b] The study and understanding of these causes constitutes a major task in astronomy.

21. Neutron stars and pulsars. Physics of black holes

As far as I know, the concept of neutron stars was first put forward[108] in 1934; it was discussed extensively for many years (see Ref. 109 and references cited there) but only theoretically. At first, attempts to observe neutron stars seemed almost hopeless, but then it was suggested that they could be found while they were still hot ($T \sim 10^6$–10^7 K) from their X-ray radiation. [The radius of a neutron star is about 10–30 km, that is, smaller than the Sun's radius (7×10^5 km) by five orders of magnitude. Therefore, at the Sun's temperature (about 6000 K) the photosphere of a neutron star would emit radiation at an intensity by ten orders of magnitude lower than that of the Sun.] In fact, neutron stars were discovered in 1967–1968 by their specific periodic radiation in the radio range; I mean here pulsars, which are at present commonly identified as neutron stars[110, 111]. There are many problems associated with the study of neutron stars and pulsars (the distinction between them should still be maintained, particularly because neutron stars do not necessarily all emit observed periodic radiation). But the same may be said about studies of stars of any type. However, neutron stars and pulsars are discussed in this book for a number of special reasons.

Firstly, the bulk of a neutron star consists of a substance with density varying between 10^{11} and 10^{15} g/cm³. We do not know properly the equation of state and properties of matter with such densities, and their study is an important task. It is especially interesting to study the superfluidity of the neutron liquid and the superconductivity of the proton liquid in neutron stars[17, 109, 112, 171]. (At densities of the order of 10^{13}–10^{15} g/cm³ the neutron substance contains a few per cent of protons and, of course, electrons; since neutrons, protons and electrons comprise degenerate fermi systems under such conditions, this mixture can be treated approximately as consisting of independent neutron, proton and electron fermi liquids.)

Secondly, not much is known about the central regions of neutron stars where densities exceed 5×10^{-14}–10^{15} g/cm³ (these densities depend on the star's mass) and, apart from nucleons and electrons, noticeable numbers of mesons and hyperons appear (that is, many species of strongly interacting particles, the hadrons) so that the equation of state becomes expecially unclear[112].

Leaving aside the hypothetical states suggested for the vicinity of singularities (cosmological, collapse), the central regions of neutron stars exhibit the highest density of matter in nature. In my view, the importance of this fact needs not be explained. [As noted in Section 19, if the fundamental length l_f exists, violations of known laws can start at a density $\rho_f \sim \hbar/cl_f^4$. Since the density in atomic nuclei is $\rho_n \sim 3\times10^{14}$ g/cm³ and no sharp anomalies of the "fundamental type" are found there, we obtain the estimate $l_f \lesssim (\rho_n c/\hbar)^{1/4} \sim 10^{-13}$ cm which is supported by even more convincing arguments; as noted above, the current estimate is $l_f \lesssim 10^{-16} - 10^{-17}$ cm. Nevertheless, the central regions of sufficiently large neutron stars can clearly have some interest for microphysics.] In addition, gravitational fields in neutron stars are also the highest found in nature (with the exception of the fields encountered in cosmological theory and gravitational collapse). Thus, deviations from GTR in stars, if they take place, should be found first in neutron stars.

Thirdly, the electrodynamics of pulsars and the mechanism by which they radiate are still not clear enough[110]. Of special interest is the structure of the crust of neutron stars[110, 111] in particular, when one takes into account the effect of magnetic fields[39, 113] which may be as high as 10^{11}–10^{13} Oe in pulsars.

Many hundreds of pulsars are currently known. But, if we ignore binary systems, only the famous pulsar PSR 0531 in the Crab Nebula emits fairly powerful optical and X-ray radiation, as well as radio waves. This is, undoubtedly, due to the young age of this pulsar which appeared with a supernova explosion in **A. D.** 1054 (the rotation period of the Crab pulsar is the shortest of the known periods, being only 0.033 s; see, however, p. 91).

The problem of neutron stars and pulsars is obviously closely linked to the problem of the mechanism of formation and explosion of supernovae. This problem has many aspects: evolution of the star before the outburst, the outburst itself, the nature of the supernova remnant, dispersion of the explosion shell, formation of some chemical elements with outbursts and so on. It should be noted that a supernova explosion can give rise not only to a neutron star but also to a black hole or a white dwarf; it can also happen that no remnant is formed, that is, the star is totally dispersed. The actual mechanism of the explosion depends primarily on the mass of the star and its chemical composition(114). A supernova explosion generally produces not only electromagnetic radiation in all frequency ranges, but also neutrinos and gravitational waves. Naturally, the study of supernovae is an extremely important branch of astromomy.

In 1971 measurements made by the Uhuru satellite revealed the existence of X-ray pulsars (sources of strictly periodic X-ray radiation). The first to be discovered were the well-known pulsars Cen X-3 (Centaurus X-3) with period 4.8 s, and Her X-1 (Hercules X-1) with period 1.2 s. About a dozen X-ray pulsars are currently known. As with radio pulsars, X-ray pulsars are clearly magnetized rotating neutron stars belonging to fairly close binary systems(115). [In principle, the role of the high-density star can be played by a white dwarf. In some cases, white dwarfs are probably radio pulsars too, but only with periods exceeding 1-3 s. The concept of quark stars has also been suggested(116). Basically, the suggestion is that at high densities exceeding the nuclear density of about 3×10^{14} g/cm^3, neutrons can be "crushed" forming "quark matter". Clearly, such stars, if they exist, are close to neutron stars or, more correctly, belong to the general type of neutron star the parameters of which (for instance, the density at the centre) can vary, depending primarily on the mass of the star.]

In binary systems plasma from the second (non-neutron) star effectively flows to the neutron star (accretion). The plasma reaching the vicinity, or the surface of the neutron star has high speed, owing to the attraction of the star. Naturally, when the plasma is stopped at the star its temperature increases greatly (to 10^7–10^8 K and more) and it emits mostly X-rays (see more on X-ray stars in Section 22).

All the above problems concerning neutron stars and pulsars involve many unclear and complicated issues. There are obvious links between these, and some primary problems of physics and astronomy. Therefore, pulsars and neutron stars will probably be studied intensely for many years to come. But if the discovery of neutron stars has just been a dream up to 1967–1968, they are now gradually becoming more or less familiar objects, although they still need getting used to.

Now the attention of novelty hunters is drawn increasingly to the search for even more exotic stars, namely, black holes. The problem, though, is not really novel; in fact it is 40 years old (we shall see later that its origin, in a sense, can be traced back to the eighteenth century) and it is associated with studies of stable configurations of the cool, "dead" stars.

Stars are heated by nuclear reactions occurring inside them, and thus emit light. The pressure gradient developed in a hot star does not allow it to contract under the effect of gravitational forces, that is, it maintains a state of quasi-equilibrium. But when the nuclear fuel burns out, the star contracts and must transform to some final (cool) state. If the star rotates slowly or hardly at all, this cooling proceeds without explosion, and if the mass of the star is less than 1.2 M_{sun} ($M_{sun} = 2\times 10^{33}$ g) then the final state is a white dwarf configuration (with radius between 10^3 and 10^4 km and mean density between 10^5 and 10^{10} g/cm^3.) The equilibrium of the star is then maintained by the "zero" pressure of the electron gas. [Such stars are typically observed in a state when they have not quite cooled down. As their surface area is small (in comparison to normal stars) this makes the temperature of the surface (photosphere) of white dwarfs typically rather high, so that they appear as "white", that is, short-wave optical radiation prevails in their spectra. But red "white dwarfs" are known, too, and the final state of any white dwarf (in the absence of accretion) is a black dwarf, that is, a completely cold and hence non-radiating high-density star. The evolution of stars, particularly the later stages, is discussed in detail in Ref. 109a.] But stars hate dying quietly and sometimes they explode when the nuclear fuel burns out, ejecting some of their matter (such explosions are observed as novae and supernovae). Perhaps (this is not yet clear) a star can disappear completely with such an explosion (that is, all its mass is ejected) but there are other, more likely, outcomes. One such possibility is that a star with $M < 1.2\ M_{sun}$ is left which evolves further into a white dwarf. Another possibility is the appearance of a neutron star, produced by the strong compression of the central regions of the initial star during explosion. If the mass of the neutron star $M < 1.2\ M_{sun}$, there are two stable equilibrium states of the cool star. Which of the two states is realized (white dwarf or neutron star) depends on the prehistory of the star (if the evolution of the star is slow, the star, of course, becomes a white dwarf). [Of course, even if $M < 1.2\ M_{sun}$ one of the two states is more energetically favourable for the star. But these states are generally divided by an enormous potential barrier. A star in such cases may jump over the white dwarf state and become a neutron star only by explosion or as a result of it.] Now, what happens to a larger star with $M > 1.2\ M_{sun}$ if it has failed to shed its shell and to get rid of a part of its mass?

Since our knowledge of the equation of state is inadequate, the maximum mass of neutron stars is still unknown. It is definitely known, however, that such a maximum mass does exist and is not more than two or three Sun's masses within the framework of GTR. Therefore, the neutron star is the only final state for stars whose mass is greater than 1.2 M_{sun}, but smaller than this maximum. [It cannot be ruled out that the maximum mass of neutron stars is smaller than the maximum mass of white dwarfs, which is (1.2–1.3) M_{sun}. Moreover, the rate of rotation of the stars is assumed to be not too high; the fate of rapidly rotating stars is substantially unclear and it is quite probable that with contraction they become unstable and break down into several stars. One can readily see that these reservations are insignificant in terms of the present discussion.] In stars of greater masses, no substance can withstand the gravitational forces and the star will contract infinitely, collapsing and becoming a black hole. A concise but clear discussion of black holes would be difficult here, particularly without using fully the GTR. Moreover, such a discussion would deviate from the general style of this book. Therefore, I shall only refer readers to the literature(88, 93b, 117, 118) and make a few remarks.

An important part in the gravitational collapse process is played by the gravitational radius

$$r_g = 2GM/c^2 \approx 3M/M_{sun} \text{ (km)} \qquad (19)$$

where M is the mass of the body, $G=6.67\times10^{-8}$ cm^3/g s^2 is the gravitational constant, and $c = 3\times10^{10}$ cm/s is the speed of light. The gravitational radius for the Sun (the mass of 2×10^{33} g), is about 3 km while the radius of its photosphere is about 7×10^5 km. For an "outside observer", that is, when the star's radiation is received far from it, the gravitational radius is the minimum radius of the surface of a contracting star since light (or any other signal) can only leave the star from distances greater than r_g. If the star's radius in its own reference system (the system linked to the star's material) is smaller than r_g then light cannot escape from the star, it is trapped by the star and "falls" down to its centre together with the material of the star.

One must not think that this effect is derived only within the framework of GTR. On the contrary, back in 1784 and 1798 Michell and Laplace noted(199) (using, of course, Newtonian mechanics and the law of gravitation) that if a star's mass is large enough light cannot leave it and for this reason, "the largest luminous bodies in the Universe may be invisible to us". The approach was correct and moreover it yielded the correct expression for the gravitational radius!

Indeed, assume that light consists of particles of mass m (in accordance with the current concepts we can take $m = \hbar\omega/c^2$ where $\hbar\omega$ is the photon energy). Such particles can escape from the distance r between it and the centre of a body with mass M to infinity if $GmM/r = mv^2/2$ where v is the radial velocity of the particle. If v is equal to the speed of light we obtain the condition $r = 2GM/c^2$ which does not include the mass m. This means that light cannot escape from distances $r < r_g = 2GM/c^2$. This calculation is of course not quite consistent, for instance, because for bodies travelling with a velocity v comparable to c, the kinetic energy is $mc^2/\sqrt{1-v^2/c^2}-mc^2$, rather than $mc^2/2$. On the other hand, if we take the energy of the particle to be mc^2 we obtain $r_g = GM/c^2$. Thus, the coincidence of the limiting Laplace radius with r_g is, to a certain extent, accidental. But it is by no means accidental that Newtonian theory is capable of describing qualitatively and sometimes quantitatively the GTR effects; one should remember that classical mechanics and the Newtonian theory of gravitation are the limiting case of GTR. [I mentioned in Section 19 above that the non-stationary state of the Universe is essentially classical in character; moreover, the laws of evolution of the Friedman's model of the Universe can be derived from Newton's theory(93). The deflection of light in the Sun's gravitational field was first noted by Soldner back in 1801. Soldner's quantitative result agrees with Einstein's predictions of 1911 (the original Soldner paper is difficult to find but his calculations are given in Ref. 119). But later, in 1915, Einstein found that the actual deflection of light had to be twice as large. As mentioned in Section 17, the observational results agree (to 1%) with Einstein's predictions of 1915, based on GTR; in 1911 GTR was not fully developed and Einstein's arguments proceeded only from the principle of equivalence which was not sufficient in this case.]

Now, let us return to black holes. This name is due to the fact that after some time τ the collapsing star dies off, and becomes invisible to an outside observer. The time τ depends on the initial conditions, the sensitivity of the observation apparatus, etc., but its order-of-magnitude estimate is

$$\tau \sim r_g/c \approx 10^{-5}\frac{M}{M_{\text{sun}}} \text{ (s)} \tag{20}$$

Thus, extinction occurs very rapidly, at least for stars with a mass $M \sim M_{\text{sun}}$, but not, for instance, for galactic nuclei and quasars with $M \sim 10^9 M_{\text{sun}}$ (if they proved to be black holes, although even for them the time τ is negligible by astronomical standards. [To avoid misunderstanding note that the time $\tau \sim r_g/c$ describes the last, relativistic stage of collapse when the star's radius r is about r_g, say r is equal or less than $3r_g$. The process of contraction to this radius may be slow but during all this period the star remains visible.] Nevertheless, a black hole cannot be said to have disappeared. First of all, one must remember that its gravitational field is fully conserved and at distances $r \gg r_g$ the gravitational potential of the star is given by the conventional Newtonian expression $\varphi = -GM/r$. Therefore, a black hole in a binary system has the same effect on the second star as a normal star. This is just where black holes are to be looked for first of all; we should try to find binary systems in which one of the stars is not radiating and its mass $M > 3M_{sun}$, and therefore it cannot be a neutron star or a dead (black) white dwarf. The task is not at all simple, but probably one black hole has already been found[120], namely, the one linked to the X-ray source Cyg X-1 (Cygnus X-1). (See also p. 91)

Of course, in this case the black hole cannot be said to be invisible. Obviously, this made the finding easier. But this apparent contradiction has to be explained. The fact is that a black hole by itself is indeed invisible (for the time $t \gg \tau$; see eq. (20)) but it is not necessarily so for the matter falling on it (accreted by it). The accreted gas is accumulated near the black hole usually in the form of a rotating disk. This gas is highly heated and emits radiation, mainly in the X-ray range. Moreover, under certain conditions the gas near the black hole can be made turbulent; as the gas approaches the region with $r \sim r_g$ magnetic fields in it increase and particles are accelerated. This gives rise to synchrotron radiation. Generally, the matter accreted by a black hole emits radiation, forming some "halo" around the hole. The resulting radiation is distinguished by its variability with a quasi-period $P \sim \tau \sim r_g/c \sim \sim 3\times10^{-5}$–$10^{-4}$ s for $M \sim (3–10)M_{sun}$. Therefore, black holes which emit radiation because of accretion are sometimes referred to as "fluctuars". Firstly, the X-ray source Cyg X-1 belongs to a rather close binary system (with period of 5–6 days); this provides for massive accretion and, almost certainly, results in X-ray radiation (the same can be said about the X-ray pulsars Cen X-3, Her X-1 and some other pulsars in which the high-density component of the binary system is a neutron star rather than a black hole; see above). Secondly, in contrast to these X-ray pulsars, the radiation emitted by Cyg X-1 has no definite period but fluctuates sharply[120]. Unfortunately, no fluctuations with a characteristic time $\tau \sim 10^{-4}$ s have yet been observed, possibly owing to the lack of suitable apparatus, and only slower fluctuations have been recorded. On the other hand, the mass of the high-density component of the binary system has been estimated as $M \sim 10\ M_{sun}$ thus supporting its identification as a black hole. However, it is not quite certain whether this system is really binary and not triple. It has also been suggested that the radiation of Cyg X-1 is due to a magnetic effect in the binary system, rather than accretion of gas by a high-density star[87]. On the whole, the nature of the Cyg X-1 source is still unclear although the most probable hypothesis for it, and some other similar sources, is that they include a black hole[120].

Of course, such a significant suggestion needs substantial proof which is still lacking.

In principle, not only a normal star with a mass $M \sim (3\text{–}50)M_{sun}$ but also a more massive object such as a quasar or a galactic nucleus can develop into a black hole. For instance, it has been suggested that our, and some other galaxies, have almost inactive nuclei at the centre which are dead quasars (that is, quasars transformed into black holes). Any remaining activity of such galactic nuclei must be associated with accretion and, in this respect, they are similar to the above "fluctuar" model, except on a much larger scale.

It has however been shown recently[121] that a large black hole ($M > 10^2$—$10^4\, M_{sun}$) cannot exist at the centre of our galaxy or of some other "quiet" galaxys; if one could exist it would attract gas and stars, giving rise to activity of the galactic nucleus, which is not observed. The hypothesis that the activity of quasars and active galaxies has a black-hole nature is quite popular, but completely lacks substantiation since an alternative, and no less probable explanation[104], exists. The elliptic galaxy M 87, which is at the same time the radio galaxy Virgo A, was studied in this respect in 1978. A large mass of comparatively low luminosity ($M \sim 5\times 10^9 M_{sun}$) is concentrated at the centre of this galaxy, a distance of 15 Mpc from us[122]. This result agrees with the hypothesis of the presence of a black hole, but falls short of a proof. Indeed, instrumental resolution available allows us only to state that the excess mass is in this region, with a size of the order of 100 pc $\approx 3\times 10^{20}$ cm. The gravitational radius for a mass of the order of $5\times 10^9 M_{sun}$ is only about 10^{15} cm. Thus, the low-luminosity mass at the centre of the galaxy M 87 may, in principle, be a dense cluster of stars[200]. Of course, the possibility of the existence of black holes in such a situation merits attention, but even for the galaxy M 87 the problem is still wide open.

The immediate future will probably see a vigorous hunt for black holes and, if they are found, an intense investigation of radiation originating in their vicinity, and other features. Such studies must take into account the rotation of black holes, and the possibility that, owing to their rapid rotation configurations with essentially different features (the so-called "naked singularities")[69 8, 117, 123] may arise. Thus, the study of black holes is now a real astrophysical problem, attracting considerable attention, although the first explicit treatment[124] of them dates back to 1939. I expect that in the immediate future the interest towards physical and astronomical processes and effects associated with black holes will keep growing.

At distances of the order of the gravitational radius the gravitational field becomes strong (the parameter $|\varphi|/c^2$ is not small and indeed cannot be used; see Section 17). Therefore, it is precisely in the vicinity of black holes that we can verify the validity of GTR in strong gravitational fields (and perhaps this may be the only possibility). For this reason, the well-substantiated discovery of at least one black hole would be especially significant in itself. The fact is that GTR allows the existence of black holes (though they must not necessarily be formed under every conditions, of course). But in some other gravitational field theories (although these are not fully developed and have been criticized[87]) black holes do not appear. Thus, the discovery of black holes would be evidence to support GTR, while not proving it. If black holes are not found (this cannot be ruled out at present) this will not be direct contradiction of GTR as it can be argued that the formation of black holes is made difficult by other competing processes (explosions of contracting masses which give rise to normal or neutron stars, and so on).

Apart from black holes, astronomers started talking comparatively recently about white holes. Assume that a star or a superstar (with mass $M \gg M_{sun}$) does not contract or collapse but, on the contrary, expands, or anticollapses. For this to happen the matter of the star must be given an initial velocity directed away from its centre. Astronomers know of one similar case, namely, the expansion of the observable Universe which left its primary state, about 10–20 billion years ago, with tremendous "initial" velocity. The hypothesis is that some of the expanding masses at this time had radii smaller than the gravitational radius, and it is only in our epoch that they "crossed the threshold" and became visible white holes observed as explosions.

Many papers have been devoted to this subject. At first, white holes were regarded as real entities in the sense that they were observable in space. On the contrary, a recent suggestion has been that white holes are unobservable.

Naturally, I do not like to write about something I do not quite understand. Unfortunately, this is the case with white holes, and therefore I will just refer readers to papers[125], the authors of which are pessimistic about the observability of white holes.

In conclusion, I shall discuss the most important recent event in the physics of black holes. In fact, it is a discovery, at the present theoretical, which is of a great general significance for physics as a whole and, perhaps, for cosmology. The very name black hole indicates that it does not emit any radiation and retains anything that gets inside it. But it has been shown[118] that if one takes into account quantum effects the above definition is inapplicable. It should be emphasized that the situation for black holes with masses of the order of the Sun's mass or larger, remains unchanged, they emit practically no radiation and the classical physics of black holes needs no corrections. But in principle black holes can have very small mass, and such holes (known as relic black holes of small mass, or just as mini-black holes) could be formed at the early (high-density) stages of the evolution of the Universe. These mini-holes emit considerable radiation, and this determines their behaviour. A non-rotating black hole with mass M, emits radiation as a black body with temperature[117, 118]

$$T = \frac{\varkappa \hbar}{2\pi c k} = \frac{c^3 \hbar}{8\pi G M k} = \frac{G M \hbar}{2\pi c r_g^2 k} \approx 10^{-7}\left(\frac{M_{\text{sun}}}{M}\right) = 10^{-7}\left(\frac{2\times 10^{33}}{M(g)}\right)\ \text{K} \tag{21}$$

Here $\varkappa = c^4/4GM = GM/r_g^2$ is the "surface gravitation"[117] (the acceleration of free fall at the surface of the black hole) and $k = 1.38\times 10^{-16}$ erg/k is the Boltzmann constant. The characteristic frequency of photons emitted by a body with temperature T is $\omega \sim kT/\hbar$; from eq. (21) we obtain $\omega \sim GM/cr_g^2$ and the characteristic time τ, $\sim 1/\omega \sim c/(GM/r_g^2)$, is the time during which a particle is accelerated to a speed of the order of c, in a gravitational field of strength, (acceleration), GM/r_g^2. Emission of photons (and particles with non-zero rest mass at sufficiently high temperatures) by a black hole is similar in character to the generation of particles in high electromagnetic fields. Photons and other particles are generated by the extremely strong gravitational field near the gravitational radius (for instance, for a black hole with mass $M = M_{\text{sun}}$, the acceleration $GM/r_g^2 \sim 10^{13} \sim 10^{12}\, g$ where $g = 981$ cm/s^2 is the acceleration of free fall at the Earth's surface).

As shown by eq. (21), a black hole with $M = M_{\text{sun}} = 2\times 10^{33}\ g$ emits radiation as a black body with temperature 10^{-7} K. Since the Universe is now filled with relict thermal radiation at a temperature of about 3 K the radiation of macroscopic black holes can be completely ignored, as it is impossible to observe. [We are not concerned here with the very distant future for open cosmological models[154]. Incidentally, discussions of the future of the Universe started only recently, although the subject is extremely fascinating.] The situation is different for the mini-holes. For instance, a hole with mass $M = 2\times 10^{15}\ g$ behaves as a black body with a temperature about 10^{11} K. The intensity of thermal radiation of a black hole is

$$\frac{dE}{dt} = \frac{10^{46} f(M)}{M^2}\ \text{(erg/s)}$$

where M is the mass of the hole in grams and $f(M)$ is a factor taking into account the emission of particles with non-zero rest mass ($f(M) \approx 1$ for $M > 10^{17}\, g$ and $f(M) \sim 10$ for $M \sim 10^{14}\, g$). The intensity of radiation of a hole with mass of $10^{14}\ g$ is about 10^{19} erg/s. The lifetime of miniholes is relatively short owing to their high radiation intensity; the characteristic lifetime for $f(M) \sim 10$ is

$$\tau \sim 10^{-27}(M(g))^2 \text{ s} \sim 10^{10}\left(\frac{M(g)}{10^{15}}\right)^3 \text{ years.}$$

Hence, only those mini-holes with $M \gtrsim 10^{15}\,g$ can exist now (if they appeared at the early stages of cosmological evolution, when the density of matter was enormously high). As no other mechanism for the formation of mini-holes has been suggested, in our time we may expect some manifestation of "evaporation" of mini-holes with masses about 10^{14}–$10^{15}\,g$; if a mini-hole of this mass has survived until now it will burn out relatively rapidly (the lifetime of a hole with mass of the order of $10^{13}\,g$ is only 10^3–10^4 years). A hole with M about $10^9\,g$ lives only for a fraction of a second, and the energy liberated during this period is $Mc^2 \sim 10^{30}$ erg. Such an explosion must be accompanied by the emission of radiation in various ranges, and could be noticed even at a great distance from the Earth. This problem has been treated theoretically in many papers (see, for instance, Ref. 126) but no observational evidence for the existence of mini-holes has yet been obtained.

The search for explosions of mini-holes is, undoubtedly quite justified, but one must bear in mind that it is still not clear whether they can be formed in any noticeable numbers. Even if GTR is valid for densities $\rho < \rho_g \sim \sim 10^{94}$ g/cm^3, (see eq. (18)), it is quite possible that mini-holes can not be formed under the specific (still unknown) conditions at the relevant stages of cosmological evolution. In addition, apparently mini-holes cannot be formed if a fundamental length $l_f \gg l_g \sim 10^{-33}$ cm exists. Indeed, the gravitational radius of a hole with mass M about $10^{15}\,g$ is as small as $r_g \sim 10^{-13}$ cm, so that the density $\rho \sim 3M/4\pi r_g^3$ is of the order of 10^{54} g/cm^3. Even here one enters the domain of microphysics. Furthermore, it would not be reasonable to introduce a gravitational radius r_g smaller than the fundamental length, l_f, so that for the minimum mass of a black hole we obtain

$$M_f \sim \frac{r_g c^2}{G} \sim \frac{l_f c^2}{G} \sim \frac{l_g c^2}{G}\left(\frac{l_f}{l_g}\right) \sim 10^{-5}\left(\frac{l_f}{l_g}\right)(g)$$

If $l_f = l_g$, then $M_f = M_g \sim 10^{-5}\,g$ and smaller holes probably cannot exist, that is, a stable particle, a maximon[3b], appears. But if $l_f \sim 10^{-17}$ cm then $M_f \sim 10^{11}\,g$. As mentioned in Section 19, the length l_f apparently corresponds to the limiting density

$$\varrho_f \sim \frac{\hbar}{c l_f^4} \sim 10^{30}\left[\frac{10^{-17}}{l_f(\text{cm})}\right]^4 \text{ (g/cm}^3\text{)}$$

If the density of a black hole during its formation cannot be higher than ρ_f, then the minimum mass of the hole is $M_{\min} \sim \rho_f r_g^3$ and $r_g = 2GM_{\min}/c^2$. Hence we obtain

$$M_{\min} \sim \frac{c^3}{\sqrt{G^3 \varrho_f}} \sim 10^{27}\left(\frac{l_f\,(\mathrm{cm})}{10^{-17}}\right)^2 (g)$$

When $l_f \sim l_g$ we have $M_{\min} \sim M_g \sim 10^{-5}$ g. But when, for instance, $l_f \sim$ $\sim 10^{-20}$ cm the minimum mass is as large as $M_{\min} \sim 10^{21}$ g and $r_{g,\,\min} =$ $= 2GM_{\min}/c^2 \sim 10^{-7}$ cm. Such calculations(127), of course, do not constitute a proof. But they show that for $l_f \gg l_g$ mini-holes with $M < 10^{15}$ g could well not be formed at all. Therefore, discovery of mini-holes with $M \lesssim 10^{15}\,g$ would, firstly, be indirect substantiation of GTR. Secondly, it would throw light on the development of the Universe at the high-density stage. Thirdly, it would put a limit on the fundamental length l_f. Unfortunately, as in many similar cases, failure to find mini-holes yields little information as it can be due to a number of causes, as mentioned above. The fact that mini-holes do not exist, does not provide any answers in itself (for instance, that $l_f \gg l_g$).

Analysis of mini-holes and their evaporation is closely linked to studies of singularities, limits of applicability of GTR, and the generation of particles in gravitational fields. As noted above, these links strongly enhance the interest in black holes, which is quite high on its own. Although the existence of black holes has not yet been proved, they already occupy an exceptional place as objects of physical and astronomical studies. If the cosmological problem is the number one problem of astronomy (many, including myself think so) then problem number two should be the problem of black holes.

*A "hyperfast" pulsar—with a period of rotation of just 1.56 milliseconds that is, rotating approximately 20 times faster than the earlier "record holder"—the pulsar in the Crab Nebula—was discovered in 1982. Apart from this discovery, and apart from other less dramatic new data on pulsars, one can also note a noticeable progress in pulsar theory, both regards their "interior" (that is, the problem of the structure of neutron stars) and as regards the analysis of models and processes in the pulsar magnetosphere.

In 1982 one observed also a new "candidate" for a black hole which is even better than the Cyg X-1 source. I refer to a binary system—the X-ray source LMC-X3 in the Large Magellanic Cloud. To be more precise, this source had been known for quite some time, but only now has it become clear that the mass of the invisible binary exceeds $6M_{\mathrm{sun}}$ which is appreciably larger than the mass limit for a neutron star.

22. Origin of cosmic rays and cosmic gamma and X-ray radiation

Cosmic rays were discovered in 1912 but it was only in 1927 that it was definitely established that highly penetrating radiation came to the Earth from space; this radiation is called cosmic rays. The nature (composition) of this radiation remained unknown for many years. We now know that cosmic rays consist of charged particles, namely protons, nuclei, electrons and positrons. To be more precise, the Earth is also bombarded by cosmic X-rays, gamma rays and, undoubtedly, by neutrinos. The term cosmic rays is now applied only to energetic charged particles of cosmic origin (this convention is further justified by the fact that in the high energy range of particle flux or energy release, charged particles dominate).

The concentration of cosmic rays (with kinetic energy, $E_{c.r.}$, of the order of, or more than, 1 GeV), near the Earth and in a considerable part of the galaxy, $N_{c.r.}$, is about 10^{-10} cm $^{-3}$, which is negligible compared to the concentration of gas particles in the galactic disk ($n \sim 1$ cm^{-3}), or even in the galactic halo ($10^{-3} \lesssim n \lesssim 10^{-2}$ cm^{-3}) or the intergalactic medium ($10^{-7} \lesssim$ $\lesssim 10^{-5}$ cm^{-3}). The energy density of cosmic rays is

$$w_{c.r.} \sim E_{c.r.} N_{c.r.} \sim 10^{-12} \text{ erg/cm}^3$$

which is no lower than the density of the internal (kinetic) gas energy

$$w_g = \frac{3}{2} knT \sim 10^{-14} - 10^{-12} \text{ erg/cm}^3$$

(here $n \lesssim 1$ cm^{-3}, $T \lesssim 10^4$ K in the disk, and $T \lesssim 10^6$ K in the halo). The energy density $w_H = H^2/8\pi$ of the magnetic field in the disk (where $H \lesssim$ $\lesssim 5\times10^{-6}$ Oe) is also no higher than $w_{c.r.}$. Thus, even in our galaxy the relativistic particles comprising cosmic rays are an important factor for dynamics and energy (this concerns, of course, the interstellar medium). Cosmic rays play an even more significant role in the envelopes of supernovae, and in radio galaxies and quasars. These findings, which have been made possible by the development of radioastronomy, are among the most significant achievements of astrophysics in the last three decades[128, 129].

The origin of cosmic rays has been discussed for several decades, but remains a sufficiently "important and interesting problem", since its great significance is obvious, and it remains a controversial subject. Until recently, three basic models have been discussed for the origin of cosmic rays—the metagalactic, halo galactic and disk galactic models. The metagalactic models assume that the bulk of cosmic rays reaching the Earth come from the metagalaxy, that is, from outside our galaxy. On the other hand, the galactic models assume that cosmic rays (with the possible exception of particles with an energy $E_{c.r.} \gtrsim 10^{17}$ eV) are produced in our galaxy, primarily by explosions of supernovae, or in the vicinity of pulsars found in the envelopes of supernovae and, possibly, by explosions of the galactic nucleus. Since 1953 my opinion has been that only galactic models are suitable. However, the metagalactic models are not easy to refute, and they continued to attract interest until recently. The metagalactic models assume that in a region around our galaxy (and maybe in the whole of the metagalaxy) the energy density of cosmic rays $w_{c.r.mg}$ is of the order of the energy density of cosmic rays in our galaxy $w_{c.r.} \sim 10^{-12}$ erg/cm^3. In contrast, models suggesting a galactic origin for the bulk of cosmic rays reaching the Earth assume that $w_{c.r.mg} \ll 10^{-12}$ erg/cm^3 (probably, as low as $w_{c.r.mg} \lesssim 10^{-15}$ erg/cm^3). Unfortunately, until now, no measurements of $w_{c.r.mg}$ seemed feasible, and a variety of estimates and indirect data had to be used.

The emergence of gamma-astronomy not only gave real hope of solving the problem by direct observation, but has already produced preliminary results which do not support metagalactic models. With galactic models, for many years discussions have focused on choosing between halo models and disk models. In the halo models cosmic rays fill a quasi-spherical or, even more flattened, large region around the galactic disk (the characteristic size of the halo is of the order of 3–10 kpc or (1–3) 10^{22} cm; recall that the distance from the Sun to the centre of the galaxy is 10 kpc). In the disk models, cosmic rays are assumed to be trapped in a disk-shaped region (the radius is about 10 kpc and the width of the disk is about 0.3–0.5 kpc). The difference between the models is reflected primarily in the predicted mean lifetime of cosmic rays in the galaxy (for protons and light nuclei this lifetime is determined by the rate of their escape from the system, that is, from the trapping region). In the halo models the lifetime is of the order of (1–3)10^8 years and in the typical disk models it is of the order of (1–3)10^6 years. Only the halo models appear consistent (the most important feature here is that the lifetime is more than, or approximately equal to, 10^8 years), but some disk models have proved very difficult to refute. This is why discussion continues, although in my opinion the question was settled in 1977 in favour of the halo models. New evidence, particularly radioastronomical data[(129, 201)], also supports these models.

There are, of course, other aspects of the problem of the origin of cosmic rays, apart from choosing a suitable model. They include the plasma effects in astrophysics[(130)], the mechanisms of particle acceleration in supernova explosions and near pulsars[(131)], solar cosmic rays and their propagation in the solar system, the chemical composition of cosmic rays, and the energy spectra of their various components (including the electron–positron component). Of especial interest is the range of superhigh energies (over 10^{17} eV).

The origin of cosmic rays of such energies is unclear[132] (some observed particles have energies as high as 10^{20} eV).

The astrophysics of cosmic rays was born after World War II and is becoming an increasingly significant part of the entire science of astrophysics. However, now this field of astrophysics is often referred to as high-energy astrophysics, and also includes questions of X-ray and gamma astronomy (high-energy neutrino astronomy should also be added).

The emergence of X-ray astronomy (if we ignore solar studies) dates back to 1962 when a powerful X-ray radiation source was accidentally and unexpectedly discovered in the Scorpio constellation (Sco X-1) in the course of rocket studies. A number of other cosmic X-ray sources (X-ray "stars") were found later; the most productive were observations from the first satellite specially designed for X-ray astronomical measurements (this satellite was launched in late 1970 by the United States from Kenya and was named *Uhuru*, which means liberty in Swahili). More than 500 X-ray "stars" are currently known, including the pulsar in the Crab Nebula, the X-ray pulsars Cen X-3 and Her X-1, a possible "fluctuar" Cyg X-1 (see Section 21), other galactic sources linked to stars, the Crab Nebula itself, other supernova envelopes, and various extragalactic sources (galaxies and quasars). A diffuse X-ray background radiation has also been recorded, that is, radiation for which no discrete sources have been identified, at any rate with the angular resolution available.

A variety of X-ray radiation mechanismus are known, namely, hot plasma bremsstrahlung, synchrotron radiation by relativistic electrons, and the scattering of radio, infrared and optical electromagnetic radiation by relativistic electrons which converts it into X-ray radiation (this process is often called the inverse Compton scattering). All these mechanisms undoubtedly contribute to the radiation flux we observe, but their contributions are different (for instance, for the Crab Nebula source the main contribution comes from synchrotron radiation, while for many other X-ray sources bremsstrahlung seems to prevail). Accretion, particularly in binary systems, clearly plays an outstanding role in the emission of high-intensity X-ray radiation. Subjects for future study are the absorption of X-rays in interstellar gas, the search for the characteristic X-ray radiation lines of various atoms and so on. On the whole, X-ray astronomy has just emerged as a major science after gathering momentum and accumulating experience for about 10 years[115, 133]. Now it is the third most important branch of astronomy, after optical astronomy and radio astronomy, in terms of classification by observational techniques or operational spectral ranges. In a short period, X-ray astronomy has yielded some first-class results (for instance, the discovery of X-ray bursts) and I am sure that more will come.

Things look different in gamma astronomy. Although the potential of gamma astronomy was first discussed[134] back in 1958 and in later publications[135], for technical reasons no reliable data have been obtained in this field for a long time. The gamma ray flux, as measured by the number of photons, is very low, although the energy flux is not so low, since each photon has a relatively high energy (the term gamma rays is applied here to electromagnetic radiation with wavelength shorter than 0.1 Å, that is, photons with energies exceeding 100 keV = 0.1 MeV). For instance, in the case of photons with energies exeeding 100 MeV one must be able to measure photon flux densities lower than 10^{-5} cm^{-2}s^{-1} (or, even better, as low as 10^{-7} cm^{-2}s^{-1}). Such fine measurements can be carried out only with devices (counters, spark chambers, etc.) having large working areas and capable of operating long enough in space. Therefore, rocket studies are not suitable for gamma astronomy although they were instrumental in the early development of X-ray astronomy.

Gamma-astronomical studies making use of high-altitude balloons and satellites have met with considerable difficulties, which have not yet been fully resolved. Nevertheless, important results have already been produced[136]. For instance, discrete sources of cosmic gamma rays have been discovered, noticeable gamma radiation from the region of the galactic disk has been reliably recorded, and an isotropic gamma radiation background of definitely metagalactic origin has been found.

Owing to lack of space I shall limit myself to noting that some gamma-astronomical observations are potentially extremely important and promising. For instance, a substantial part of gamma radiation with energies exceeding 50–100 MeV must be generated by the proton–nuclear component of cosmic rays in the interstellar and intergalactic media. The protons and nuclei of cosmic rays collide with protons and nuclei of the interstellar and intergalactic gas giving rise to neutral pions among other particles. Pions decay immediately into two gamma photons (the mean lifetime of a pion is 0.84×10^{-16} s), each with an energy $m_{\pi^0}c^2/2 = 67.5$ MeV (pions are assumed to be at rest). Gamma photons are emitted also in the decay of the Σ^0 hyperon ($\Sigma^0 \rightarrow \Lambda+\gamma$) and in the decay of some mesons and hyperons giving rise to neutral pions ($K^{\pm} \rightarrow \pi^{\pm}+\pi^0$, $\Lambda \rightarrow n+\pi^0$, etc.). Such gamma rays of "nuclear" origin are distinguished by their spectrum (generally, their energy is higher than 30–50 MeV) and, in principle, they can be identified by their difference from gamma rays produced by other processes, for instance, as a result of bremsstrahlung of relativistic electrons. The flux of "nuclear" gamma rays is proportional to the intensity of the cosmic rays which generate them, and this relationship makes it possible to determine this intensity in regions far from the Earth, at the galactic centre, in radio galaxies and so on.

Until the emergence of gamma astronomy all data on the main, proton and nuclear, component of cosmic rays in regions far from the Earth were obtained either by extrapolation from results on cosmic rays in the vicinity of the Earth, or from estimations involving additional assumptions often quite reasonable, based on radioastronomical measurements [information on relativistic electrons in the radio-emitting regions is deduced from radio observations (also with the use of additional assumptions), but at least in a more direct way (128, 129)].

More or less direct determinations of the intensity (and energy density) of protons and nuclei on cosmic rays far from the Earth will be highly important. For instance, only they can be expected to finally resolve the protracted controversy about the galactic or metagalactic origin of cosmic rays(129, 201). In metagalactic models, the energy density of cosmic rays in the relatively small galaxies nearest to us, the Magellanic Clouds, must be the same as in our galaxy, and the surrounding regions, that is, about 10^{-12} erg/cm^3. Since we know the amount of gas in the Magellanic Clouds we can estimate that at the Earth the flux of gamma radiation emitted from them must be about 3×10^{-7} cm^{-2} s^{-1} (for photon energies exceeding 100 MeV). If a lower gamma flux is measured the metagalactic models will be completely refuted. If measured fluxes are in this range, or higher, the situation will remain unclear, as in this case gamma radiation can be produced by cosmic rays originating within the Magellanic Clouds[218].

Gamma astronomical studies of the Magellanic Clouds will only be possible in the (hopefully not too distant) future. But measured fluxes of gamma rays of galactic origin coming from the direction of the galactic anticentre show that the energy density of cosmic rays decreases noticeably with increasing distance from the solar system(136, 201) (in the direction towards the anticentre). This result is quite natural for galactic models but disagrees with metagalactic models (see above).

Apart from gamma rays produced by the decay of pions, it would be very interesting to study gamma rays of different origins, for instance those generated by relativistic electrons, in nuclear transitions(137), and so on. Thus, gamma astronomy has (potentially) very wide scope. Jo conclude we shall discuss one more major discovery of gamma astronomy.

In the 1960s the USA launched four satellites of the Vela series which were designed for inspection, in accordance with the treaty prohibiting nuclear explosions in outer space, and carrying gamma ray detectors for this purpose. No nuclear explosions were detected, but in the period from July 1969 to July 1972 the detectors recorded 16 bursts of gamma radiation(138a) lasting from fractions of a second to tens of seconds. It was especially significant that the bursts were recorded simultaneously by several Vela satellites at large distances from one another. Thus, it could be ruled out that a burst was an artefact caused by an instrumental fault in one sateillite. Later, a check was made of records from other satellites with appropriate equipment which were operating through out the same period[138b]; the check revealed some of the bursts recorded by Vela satellites (one could hardly expect that all satellites recorded all bursts since detectors do not function all the time, a satellite may be in the Earth's shadow, and so on). One should not think that different satellites record bursts strictly simultaneously. This cannot happen as the speed of light (that is, gamma photons) is finite, while the distance between satellites may be large (for instance, the Vela satellites are at a distance of about 120 000 km from the earth's centre, so that the distance between them may be as large as 240 000 km and the maximum delay between bursts may be almost a second, while bursts are recorded to hundredths of a second). Incidentally, the fact that these bursts do not originate in the Sun or on the Earth has been established on the basis of the measured delays and the relative positions of the satellites.

The amount of data on gamma bursts is not insignificant, but so far no "unusual" visible objects (for instance, remnants of supernova explosions) have been found in the regions from which the bursts came. A possible exception is the powerful and peculiar gamma burst recorded on March 5, 1979 (see Refs. 202, 203 and papers cited there). The source of this burst is possibly a supernova remnant (apparently a neutron star) in the Large Magellanic Cloud. This assumption is difficult to substantiate in terms of the energy balance but still it seems permissible[203]. Leaving aside this burst, one should note that the intensity of the gamma bursts (they were observed in the energy range from 0.1 to 1.5 MeV and some in the X-ray range) is quite high, so that they must have originated in some powerful cosmic explosions. Indeed, for the burst which was studied best during initial observations, the total energy received in approximately 80 s was $\phi \sim 5\times10^{-4}$ erg/cm^2 (this was the duration τ of the burst). If the source of the radiation is in the galaxy at a distance of, say, $R \sim 100$ pc $\sim 3\times10^{20}$ cm then the total energy released at the source is $W \sim 4\pi R^2 \sim 10^{39}$ ergs and the respective power is $L \sim W/\tau \sim 10^{37}$ erg/s. If the source is in nearby galaxies (for example, it is a supernova) and $R \sim 3$ Mpc $\sim 10^{25}$ cm we have $W \sim 10^{48}$ ergs and $L \sim 10^{46}$ erg/s. Finally for the farthest possible sources (such as collapsing galactic nuclei) and $R \sim 10^{28}$ cm we have $W \sim 10^{54}$ ergs $\sim M_{sun}c^2$ and $L \sim 10^{52}$ erg/s. Note that the total intensity of the solar electromagnetic radiation (luminosity) is 3.86×10^{33} erg/s; this shows that the sources of gamma bursts are very powerful even by cosmic standards.

A number of reports on gamma bursts have been published recently[111, 202–204]. New results[204] suggest sufficiently convincingly that gamma bursts originate in our galaxy and are somehow associated with neutron stars.

In recent years the study of gamma bursts has been in a sense rivalled by the discovery of the much more frequent X-ray bursts[133] first reported in 1975. The sources of these X-ray bursts are, undoubtedly, in our galaxy since they are concentrated near the galactic plane. These bursts are possibly associated with the irregular accretion of plasma by neutron stars (and, perhaps, white dwarfs and black holes), with a contribution from fusion reactions in matter accreted by neutron stars[133]. One cannot rule out the possibility that gamma bursts have the same nature as X-ray bursts, and are produced in those cases where the energy of the emitted photons is high enough to be in the gamma range (note that gamma bursts are recorded in the soft gamma ray range, primarily at energies of hundreds of kilo-electron-volts). This assumption has however, been met with some criticism, at least, with respect to most sources of X-ray bursts (a few tens of them are currently known). Apart from the discovery of X-ray pulsars, the discovery of X-ray and gamma bursts can be regarded as the major events in observational astronomy since the discovery of radio pulsars in 1967–1968.

If the reader has not been fascinated by the prospects of high-energy astrophysics, this may only be due to my failure to depict convincingly the impressive achievements and significance of this new branch of astronomy; but I hope this is not the case.

23. Neutrino astronomy

Pauli predicted the existence of the neutrino in 1931. In took 25 years (a longish time for our fast-moving world) to record neutrinos in the vicinity of nuclear reactors. Then it was a natural question to ask if we could detect neutrinos of extraterrestrial origin.

Since nuclear reactions are the sources of energy of all stars neutrinos must clearly be emitted by all stars. First of all, one should, of course, consider the Sun. The distance betweeen the Sun and the Earth is 1.5×10^{13} cm while the distance to the nearest stars is of the order of 4×10^{18} cm, so the flux of solar neutrinos must be 10^{11} times the flux from the nearest stars, other conditions being equal. Attempts to detect solar neutrinos by means of the nuclear reactions $^{27}\mathrm{Cl}+\nu_e \rightarrow {}^{37}\mathrm{Ar}+\mathrm{e}^-$ (here ν_e is the electron neutrino and e^- is the electron) started more than ten years ago[128, 139], but did not yield any positive results for a long time. The neutrino flux was expected to be of the order of tens of solar neutrino units (SNU). For a flux of 1 SNU, on average one neutrino per second is captured by 10^{36} of ^{37}Cl nuclei. According to available data[140], the flux of solar neutrinos recorded by the chlorine detector must be 4.7 SNU (for the so-called standard models of the Sun). Other estimates also have been reported[205], for instance, 6 ± 2 SNU. But the recorded neutrino flux was much lower and could even have been zero, within the range of measurement errors. This disagreement stimulated great interest and gave rise to a variety of hypotheses and speculations. However, this question actually never gave ground for any far-reaching conclusions since the chlorine detector effectively records only neutrinos of fairly high energy (higher than 0.81 MeV), produced mainly in the radioactive decay of ^{8}B nuclei, which is a rare reaction in the Sun. The flux of such neutrinos is very sensitive to the temperature deep in the Sun and, in general, greatly depends on the chosen model of the Sun. However, the chlorine detector should not record a flux lower than approximately 0.5–1 SNU and, especially, lower than 0.25 SNU, even with far-reaching assumptions about the structure of the Sun. If the flux were lower than the above limit we would have to assume that the Sun is non-stationary (in particular, that the temperature in the Sun's core fluctuates and is currently lower than in the steady-state models of the Sun), or that neutrinos are unstable (that is, a noticeable decay of neutrinos occurs during the eight minutes it takes them to travel from the Sun to the Earth). Even more radical suggestions have also been made[140, 141].

In a way, the most interesting outcome would be if some essentially new phenomena were responsible for the results on solar neutrino flux, but this is hardly the case. Indeed, recent measurement of solar neutrino flux[205] have yielded a value of 2.2 ± 0.3 SNU, while another reported value[206] is 1.8 ± 0.4 SNU. Of course, such accuracy is unsatisfactory, but still the results are lower than the "standard" flux of 4.7 SNU by a factor of only two or three. The reasons for such disagreement may not be very significant. However, the problem is yet to be resolved.

An especially interesting effect is neutrino oscillations (see Refs. 141, 205 and Section 14). If they existed then under certain conditions (depending primarily on the difference between the masses of neutrinos of different species, that is, ν_e ν_μ and ν_τ) the recorded neutrino flux would have to be lower by a factor of three than the flux predicted without taking into account the oscillations[205]. As noted above, the predicted results and experimental data do indeed disagree by approximately a factor of three. However, it is too early to say that it is precisely neutrino oscillations that are responsible for this disagreement.

The work with chlorine detectors will be continued but it must be supplemented by measurements with other detectors, primarily, ^{7}Li and, especially,

^{71}Ga. The isotope ^{71}Ga reacts with neutrinos whose energy is just over 0.23 MeV and gives rise to ^{71}Ge. Therefore, a gallium detector will be capable of recording the bulk of solar neutrinos produced in the reaction $p+p \rightarrow d+e^{+}+\nu_e$ with energies up to 0.42 MeV. The flux of these neutrinos is determined, to a good approximation, by the luminosity of the Sun and therefore it is independent of the model of the Sun (under an assumption that the flux is stationary). Techniques for separating germanium from gallium are available and the use of a gallium detector (whose mass must be 20–40 tons) opens up significant prospects(140, 206).

The emergence of neutrino astronomy is a major event as detection of neutrinos is the only known method for probing the central regions of stars. But one can scarcely hope for detection of neutrinos produced in "normal" stars (apart from the Sun) in the foreseeable future. Things are different with supernova explosions and the formation of neutron stars, which may give rise to high-intensity neutrino fluxes(128, 139, 140, 206) (these events are not the same since a supernova explosion may also give rise to a white dwarf, or a collapsed object known as a black hole, or result in the complete disappearance of the star).The same may be said about such—still somewhat hypothetical—events as the collapse of supermassive stars (including galactic nuclei).

Finally, it would be an extremely significant result if we managed to detect neutrinos produced at early stages of the evolution of the universe(93b, 139, 140, 190). Unfortunately, prospects in this respect are not very encouraging since first the sensitivity of the detectors available must be increased by a few orders of magnitude. But, as we know from the history of physics and astronomy, where pessimistic predictions most often prove to be false, it is precisely because predictions about measurement techniques have been wrong. In addition, current estimates of the flux of the "cosmological" neutrinos may prove to be too low.

Apart from the subjects of neutrino astronomy discussed above, studies of high-energy neutrinos(72, 140, 142, 207) (see Section 13) are currently attracting considerable attention. These neutrinos are produced practically only by the proton and nuclear components of cosmic rays. In this respect they are similar to the gamma photons produced by the decay of neutral pions (see Section 22). But neutrino-producing cosmic rays are of much higher energy, and generate neutrinos with energies exceeding 10^3 GeV. Studies are planned (primarily the DUMAND project, according to which neutrino-produced showers will be optically recorded in deep water) which will probably make it possible to record neutrinos from quasars and active galactic nuclei(141, 207). Perhaps it will be precisely such studies that will reveal whether the quasar core is a massive black hole or a magnetoid(297).

Thus, neutrino astronomy is an up-and-coming research field promising to bring forth valuable results and, possibly, new discoveries.

24. The present stage in the development of astronomy

Several first-rate astronomical discoveries have been made in the last 20 years, including quasars, relict thermal radiation, X-ray "stars", the maser effect in space with OH, H_2O and other molecules, pulsars, X-ray and gamma bursts, not to mention many other significant achievements on a somewhat lesser scale. Recent advances in astronomy look even more impressive if we add to them some of the achievements in space research such as lunar and planetary studies.

Development of different fields of science is qualitatively uneven. As for astronomy, one may say that after World War II it entered a period of spectacular growth. I wrote about it on several occasions[143, 208] (as did many other authors) but still I feel that the astrophysical part of this book should end with some remarks of a general nature (see also Section 25).

Firstly, the advances in astronomy are, undoubtedly, due to the development of physics and space technology, which made it possible to design and work with equipment of fantastic sensitivity, and in some cases to launch it beyond the Earth's atmosphere. [To illustrate this point about the sensitivity of astronomical apparatus I can describe something which impressed me some years ago, although by that time I had worked in radioastronomy for many years. This happened at a small exhibition organized by the radioastronomical observatory near Cambridge (Britain). Visitors were invited to take small sheets of plain white paper lying on a table. A visitor took a sheet, turned it over and saw the following words: "Taking this sheet from the table you have spent an amount of energy which exceeds all the energy received by all radiotelescopes in the world throughout the entire history of radioastronomy". The flux density or, more exactly, the spectral flux density is typically measured in radioastronomy in units of 10^{-23} erg/cm^2 s Hz $=$ 10^{-26} W/m^2 Hz. A flux of such density delivers an energy 3×10^4 ergs $=$ 3×10^{-3} J in a spectral band with a width of 10^{10} Hz to an area of 1 km^2 $= 10^{10}$ cm^2 per 1 year ($= 3\times10^7$ s). Available apparatus can detect radiation from sources with a flux of one unit (or even lower by two or three orders of magnitude). But typical radioastronomical sources produce fluxes about ten times greater than that, and the total number of such sources is just a few hundreds. These figures demonstrate the validity of the above example and illustrate vividly the amazing sensitivity of radioastronomical equipment.]

Secondly, changes taking place in astronomy consist primarily in the transition from optical astronomy to astronomy operating throughout the entire electromagnetic spectrum. Thirdly, the latest astronomical advances, remarkable as they are, have not yet necessitated any revision of the fundamentals of physics, as they may be explained within the framework of existing physical concepts and laws.

Not everybody would agree with the second and third conclusions. Some would say that the major feature of the current stage in the development of astronomy is the emergence of new concepts or revolutions in ideas. However, recent astronomical discoveries, significant as they are, are no more profound or far-reaching than the discovery of the expansion of the Universe and the evaluation of its characteristic dimensions (the time $T \sim 10^{10}$ years and the distance $R \sim cT \sim 10^{28}$ cm). All this was done basically in the 1920s. As for the claims that astronomy has recently given rise to "new physics", it has been mentioned in Section 20 that opinions differ in this respect and arguments refuting this point of view have been presented.

What will happen next? What are the trends in the development of astronomy? In such matters predictions are a risky business. But it is better to make an error than to keep cautious silence. Hence, I shall venture some predictions, which are far from unorthodox.

It can be expected that the present stage in the development of astronomy (the second astronomical revolution; see Section 25) will be completed fairly soon (say, by 1990), astronomy will envelope the entire electromagnetic spectrum, and all the discoveries that are, so to speak, within easy reach, will be made. The following period should apparently be a less turbulent one (we are talking here only about studies of far-away objects, leaving aside planetary studies and the fascinating problem of extraterrestrial civilizations[(144)]). In other words, the pioneering period will end and the curve depicting the development of astrophysics will become less steep, if only for some time. However, it should be noted that astronomy has a rich development potential, depending on the progress of neutrino astronomy and the astronomy of gravitational waves, as well as the construction of gigantic radiotelescopes in outer space.

Finally, let us consider the principal (at least for physicists) question of whether the development of astronomy can lead to changes in fundamental physical concepts, which seem to be greatly desired by some astronomers. Among such possible changes I can mention the introduction of scalar fields into the relativistic gravitational theory, violation of conservation of baryon and lepton charges, deviations from existing physical laws at high densities inside or near enormous masses (galactic nuclei, quasars and neutron stars), variation of physical constants with time, and so on. [In this connection, it is especially interesting to ask whether the gravitational constant G depends on time[146]. The Universe is non-stationary (it expands) and, at the same time, its dynamics are determined by gravitational interaction. Therefore, an assumption about the time dependence of the gravitational constant does not seem totally groundless, although it is not at all necessary either for a logical treatment, or for explaining experimental and observational data. But this matter can only be resolved by experiment. The general theory of relativity can serve as a foundation for cosmological studies only if $(1/G)(|dG/dt|)$ is less than 10^{-11} year^{-1}. Available measurement techniques are still insufficiently accurate to determine this limit reliably, but are close to it. The value of dG/dt or its upper limit can be found from measurements of the orbits of the artificial and natural planets, and their satellites, or by using high-sensitivity gravimeters[146c]. Another, no less significant, issue discussed widely recently is the suggestion that protons are unstable (see Sections 14); the lower limit of their lifetime is $10^{31}-10^{32}$ years.]

The search for new fundamental ideas and concepts in astronomy (including cosmology) is, of course, of the utmost significance, but essentially the problem defies prediction. Thus, the "principal" question stated above remains, in fact, unanswered. I can only note that, speaking for myself, I would not be surprised at all if new physics were needed in astronomy (and stimulated by astronomy) only in the vicinity of the classical singularities, that is, if it were essential only for cosmology, and for explaining the final stage of gravitational collapse, as well as for predicting the far future of the Universe[154] (actually, I tend to believe that it will be so).

Another aspect of this problem is as noted in Section 14, that the study of the early stages of cosmological evolution is extremely important for physics or, more precisely, for high-energy physics.

It well may happen otherwise, and astronomical discoveries may make a wider and more varied contribution to the fundamental ideas and concepts of physics. What is emphasized here is that it is not necessarily so, and no historical associations, general concepts or available knowledge can provide an answer to this question.

Concluding remarks

25. Some notes on the development of science

Science occupies a significant place in the world today. For instance, the jobs of millions of men and women are to a varying degree linked to scientific activities. It is only natural, therefore, that science itself, its development, and various features are analysed in numerous papers, reports and studies by people concerned with science policy studies, the methodology of science and the history of science. As far as I can judge though, active physicists and astronomers (and, probably, scientists working in other fields) are relatively uninterested in the science development and policy studies, and the history of science. This is understandable since any specific problem, say, in solid state physics, seems to be separate from the general development of world science and its history for the scientist working on it. But anybody who is concerned with the future of science, and trends in its development on the world scale or on a smaller scale (say, in one country) is bound to tackle some issues of the history and development of science. And, of course, many such issues are fascinating just in themselves. One such widely discussed issue is the structure of revolutions in science, and the very concept of scientific revolution.

The significance of this last issue seems, at the least, doubtful as debates on the meaning of scientific revolution are largely scholastic or semantic. However, I was very disturbed by the well-known book(209) by Kuhn *The structure of scientific revolutions*, and thus somehow got involved in the discussion of them(147, 208). Amongst other issues, the discussion touched upon the character and meaning of the revolution in astronomy (see Section 24).

In the last Russian edition of this book (1980), as in earlier editions, I linked the first astronomical revolution to Galileo's work (the introduction of the telescope and the resulting discoveries) and the second revolution to the transition from optical astronomy to astronomy operating throughout the entire electromagnetic spectrum. This approach was justly criticized. Indeed, is the introduction of the telescope more significant than the work of Copernicus, which is a typical example of scientific revolution? Can one not classify as astronomical revolution the discovery of the non-stationarity of the universe, and the resulting development of extragalactic astronomy? These questions made me review the issue of scientific revolutions(208) and I shall here explain my opinions briefly.

Kuhn writes that, for him, a revolution is a type of change including a certain reconstruction of the guidelines governing a group. This change is not necessarily great, and may not seem revolutionary to those who are outside a separate (closed) community which may consist of as few as 25 persons[209].

This definition reminds one of a tempest in a teapot. As is usually the case in questions of terminology, one cannot forbid or refute most definitions. Between the extremes of a tempest in an ocean and a tempest in a teapot there are numerous cases differing in scale (tempests in gulfs, lakes, ponds, swimming pools, and so on). In a similar way, revolutions in science can be classified, as is sometimes done, into global, local, minor, etc., revolutions.

Finally, any breakthrough or "reconstruction of guidelines"[209] may be termed a scientific revolution. In my opinion such an approach is unreasonable. A genuine revolution implies a sharp change or a radical transformation. It would be natural to define the scientific revolution as a profound change in all natural sciences, or in a major part of them (physics, astronomy or biology, for instance). The fitting of a revolution to a hierarchy of Procrustean beds (global, local, minor, etc.) is hardly worthwhile. A worthwhile task would be the detailed analysis of historical and current situations in science.

Now I would like to say a few words about my opinions on revolutions in physics and astronomy. Only two revolutions have taken place in astronomy, if one leaves aside its emergence in antiquity (the astronomical system of Hipparchus and Ptolemy). The first revolution occurred in the sixteenth and seventeenth centuries and is linked primarily to the names of Copernicus and Galileo. The revolution consisted in the transition from the geocentric system to the heliocentric system (see, for instance, Ref. 87b) and Galileo's reform of observational techniques (the introduction of telescope) resulting in a number of brilliant discoveries.

The second astronomical revolution took place (in fact, is still taking place) in this century. It, too, has two aspects. On the one hand, the second revolution is due to the discovery of non-stationarity of the Universe and the emergence of extragalactic astronomy. On the other hand, it consists of the transformation of optical astronomy into astronomy operating throughout the entire electromagnetic spectrum. As with Galileo's innovations, this transformation resulted in a variety of remarkable discoveries which have been briefly discussed in this book.

This approach (or classification) makes each revolution somewhat longer, but seems sufficiently consistent and, most importantly, agrees with historical fact.

If one turns to physics one can also distinguish two revolutions (again leaving aside antiquity). The first revolution consisted of the development fo classical mechanics (Copernicus, Galileo, Kepler and, especially, Newton) and is essentially the foundation of what is now known as classical physics. The progress due to the work of Faraday, Maxwell and the other creators of electrodynamics cannot be overestimated, but in the last century electrodynamics did not seem to disagree with classical mechanics. Therefore, the second revolution in physics should be linked to the development of the theory of relativity and quantum mechanics. Dates in such cases are usually controversial, but in my opinion the second revolution should be dated as 1900–1930.

The above opinions lead to a natural question about the current stage in physics. Are we between revolutions, or has the third revolution already begun? Could the character of the development of physics, and science in general, have altered in such a way that the above concept of scientific revolution is no longer applicable? Such questions may have interesting answers and I have attempted to answer them elsewhere[(208)]. But these issues are irrelevant to the framework of this book. What seems more relevant here is a discussion of the rate of growth of science.

The rate of growth of science as a whole, and in the individual sciences (physics, mathematics, biology, etc.) has been fairly constant for about 300 years, amounting to 5–7% a year. This means that various scientific "indices" or "products", such as the number of scientists, the number of scientific papers, the number of journals, etc., has increased according to the exponential function

$$y_i(t) = y_i(0)e^{t/T_i} \tag{22}$$

where $y_i(0)$ is the product (say, the number of physical journals) at the moment $t = 0$ (the reference point in time), $y_i(t)$ the product at the moment t, and T_i is the time interval during which the product has grown e (≈ 2.72) times. For instance, a 7% growth per year corresponds to a characteristic time $T_i = 15$ years; in 15 years the products increase 2.72 times, in 30 years 7.4 times, in 60 years approximately 50 times, and in 120 years 2500 times. The fact that the exponential law (22) of development is applicable in many (though in by no means all) cases is quite natural; the growth dy_i of the product during a short time interval dt close to the moment t is proportional to the product at the moment t, that is

$$dy_i = \frac{y_i(t)}{T_i}\,dt$$

In the absence of limiting factors this is exactly what happens, for instance, with the number of scientists and research papers—the larger the number of scientists, the larger the number of people they train with a corresponding increase in the number of papers. Though the law of exponential growth is well known and quite simple, its consequences are somehow difficult to grasp in their entirety. Since in developed countries the "lifetime" of one generation is about 30 years (this is about the average age difference between parents and children) the exponential growth with $T_i = 15$ years means that during the lifetime of one generation of scientists their number increases 7.4 times and that the scientific product produced throughout the entire history of humanity is smaller by a factor of $[y_i(2T_i) - y_i(0)]/y_i(0) = 6.4$ than the product produced in the last period of 30 years.

Another vivid illustration is that about 90% of all scientists who have ever lived on Earth live at present. [I used this well-known illustration elsewhere(147) and a curious thing happened. Either the copy editor or the proof-reader could not believe that 90% of the scientists live at present and put the word live in quotation marks! This shows that it is indeed not easy to appreciate the implications of exponential growth; see also Ref. 155.] In 1913 the number of scientists in Russia was below 12 000; by the middle of the 1970s the number of scientists in the USSR was about 1 200 000, that is, in 60 years the number has increased 100 times. If this growth rate continues for another 60 years one-third of the population of the USSR will be scientists.

This is clearly impossible and sooner or later the growth of science or, at least, the growth of the number of scientists, and some other indices, must slow down or even stop altogether. In the developed countries the saturation effect in the development of science has already manifested itself to a certain extent. On the other hand, the level of demands made on science by technology and society in general is by no means decreasing. This gives rise to a contradiction between the need to slow down the growth of the number of scientists and the demands made on science.

The problem can be resolved in only one way, namely, by improving the efficiency of research work. But going along this way involves fundamental difficulties. Indeed, prospects for improving labour productivity in industry or agriculture are almost infinite or, at least, very great. But for creative work, in particular scientific creative work, such prospects hardly exist.

Of course, much can still be done in practice by improving the conditions of research work and, more significantly, by extensively using computer technology. But still the "bottleneck" in science will be determined by human qualities. One cannot, of course, expect that within the next decades human qualities will change (that is, human capabilities will improve). However, this statement should be qualified. Some people have been known to possess an extraordinary memory, or to perform fairly complicated mental mathematical calculations very rapidly. Since they are not little green men but human beings who are quite ordinary in other respects, their extraordinary capabilities evidence enormous reserve capacities of the human brain. This opinion is supported by other data. The existence of the considerable reserve capacities of the brain seems to be a natural result of biological evolution; as with any other body organ, the brain must operate reliably, and thus possess a significant "safety margin". But in our time it seems quite reasonable to attempt to tap the reserve capacity of the brain to improve the efficiency of creative work. Although I am not an expert, and can hardly evaluate the chances of success here, I still believe that it is one of the most important and interesting problems of biology.

In conclusion, I shall note one psychological effect of the rapid growth of the number of scientists. Because of such growth the average scientist is relatively young. I do not know the exact figures but the average age of physicists is probably between 35 and 40 years. For a 35-year-old person everything that happened 30 or more years ago seems prehistoric, and in science many things that happened 15 or more years ago, that is, before the start of his or her active professional work, seem ancient history too (some qualifying remarks would be in order here but I hope my reasoning is clear enough without them). Such a situation gives rise to an overestimate of the development rate of science, which is widely current in the scientific community. A young scientist feels that 10, 15 and, of course, 25 years are very long times, not only in terms of a human lifetime, but for science as well. Such a feeling does not always reflect the truth, however. Suffice it to note that the special theory of relativity is over 75 years old, the general theory of relativity is over 60 years old, non relativistic quantum mechanics emerged over 50 years ago, superconductivity was discovered in 1911, and cosmic rays were discovered in 1912. But after seven decades both superconductivity and cosmic rays still attract a lot of attention, and various aspects of their study are included in the present list of the most important and interesting problems of modern physics and astrophysics (see Section 2 and 22). Moreover, the history of these fields, of which I have a sufficient knowledge, shows that it took 25 and even 45 years to solve some of their problems (for instance the microscopic mechanism of superconductivity was not understood until 1957).

What are the conclusions? The only one is that we cannot expect breakthroughs in science every year or even every decade. For instance, macrophysics and astrophysics have made, of course, considerable progress in the last decade but no fundamental breakthroughs. As for microphysics, no definite judgement can be made as yet. Progress in microphysics has been so great that, perhaps, the 1970s will mark an exceptional stage in the history of microphysics, for example, as the period in which the fundamentals of the unified field theory were developed (it is still too early to make any final conclusions as to the scope and extent of success of attempts to unify the theories of the weak, strong, and electromagnetic interactions, let alone gravitational interaction). But even from this viewpoint the 1950, and '60s can scarcely claim a similar role, although much was done in this period.

What will happen before January the 1st, 2001—the beginning of the twenty-first century? Less than 20 years are left. For schoolchildren it is a long time, but for people who have worked for something like 40 years a period of 20 years does not seem too long—just recall what has happened since 1960 (was physics at that time significantly different from physics today?).

I do not have much chance of seeing the beginning of the next century, and my chance of being able to evaluate the state of science then will be even less. But I hope that the majority of readers of this book will meet the twenty-first century in the prime of life, and I would like them to think then about updating this list of the "most interesting and important" problems. I would not be too surprised if a good half of the problems discussed in this book reappeared in a list compiled in 2001.

26. In lieu of conclusion

This book touches upon many subjects and problems. It is not possible or necessary to end it with any specific conclusions. I shall only make a few general remarks, addressed to the so-called "unsophisticated" reader.

The history of science abounds with wrong predictions. A good illustration may be found in a speech delivered by Rutherford, the discoverer of atomic nuclei and nuclear transformations, to the annual meeting of the British Association for the Advancement of Science on September 11, 1933. Rutherford said in his speech (which was extensively reported in newspapers) that "all talk of using nuclear energy is moonshine". His opinion was shared by other people and he was, in a sense, quite right because in 1933 no clear way of using nuclear energy could be seen. However, in barely five years things changed radically as uranium fission was discovered and in nine years (in 1942) the first uranium pile was put into operation.

This and similar examples may give rise to a profound distrust of any planning or prognostication in science. Specifically, doubts are raised about the very idea of discussing some "especially important" but still completely unsolved problems. In this connection I would like to emphasize the following. In many instances (perhaps typically) it is impossible to plan or predict the course of fundamental scientific studies, in the sense of specifying deadlines and so on. For instance, when will high-temperature superconductivity be discovered? As follows from the discussion of this problem in Section 2, I would answer the question in this way: perhaps high-temperature superconductivity has already been discovered in some laboratory (but the discovery has not been reported yet), perhaps it will be discovered tomorrow, and maybe such a phenomenon cannot occur at all and hence will never be discovered. In other words, time limits for making discoveries or solving problems in science can be only very unstable and vague and it is better not to talk about them at all.

But one can, of course, talk about the problem itself! For instance, when the nuclear mass defect was discovered it became clear that vast energy was stored in nuclei. This resulted in the emergence of the problem of nuclear (atomic) energy in the 1920s. Naturally, this problem had to appear in any reasonable list of the "most important" physical problems until the early 1940s when it was solved after 20 years of work. Generally speaking, the scientific problems themselves seem to be rather stable.

Thus, there is nothing to say against planning and prognostication in fundamental research, if they imply the formulation of current problems, the tentative evaluation of their potential importance, and so on, rather than fixing deadlines (of course, I do not mean stipulating dates for putting into operation scientific apparatus, etc.).

However, any selection of the "especially important and interesting" problems tends to be rather arbitrary. It is quite clear that different "important" problems are neither equivalent nor easy to compare, and their selection changes with time. For instance, if at least one superconductor with a critical temperature of the order of room temperature were produced, and the factors contributing to it were understood, the problem of high-temperature superconductivity would probably be eliminated from the present list of "important" problems. A problem should also be eliminated from the list if the answer to it is proved to be negative. For instance, this would be true if it could be shown that superconductors with high critical temperatures, or long-lived superheavy nuclei do not exist.

To avoid misunderstanding, I should once more emphasize that problems not included in this list must, of course, be studied too. Apart from the fact that no well-defined boundaries exist between various physical and technological research and development projects, one has to bear in mind how new "especially important" problems tend to emerge. They mostly originate, as do discoveries, from the "grass-roots" problems, just as geniuses are born to normal parents. In the 1930s it could hardly be said that studies of the luminescence of liquids under the effect of gamma rays were especially important, and yet these studies resulted in the discovery of the Cerenkov effect. The same may be said of the Mőssbauer effect, some recent astronomical discoveries, for instance the discovery of pulsars, and so on.

In other words, many remarkable discoveries and achievements in science are unforeseen and unexpected.

Generally, while it is only natural and reasonable to concentrate efforts on solving current "especially important" problems, other fields of research must not be neglected and physics and astrophysics should develop harmoniously together. [A harmonious development is, of course, very difficult to achieve. In the USA special physical and astronomical commissions have been established by the National Academy of Sciences to analyse this problem. The reports of these commissions(148) present interesting material on the prospects of development of physics in the USA; see also Ref. 149.]

As emphasized above, identification of the "especially important and interesting problems" is in general rather arbitrary, insufficiently clear and ambiguous and, of course, this is true to an even greater extent with respect to particular cases or opinions of individuals or small research groups. If, for instance, a physicist has discovered (experimentally or theoretically) a new effect or a new measurement technique, this effect or technique becomes the most important and interesting one for him for a time. It typically does not matter whether this research field is fashionable or not, or whether it appears on some selected "lists". For instance, though for a long time I have been advocating selection and identification of important problems and fields, I do not work exclusively (or even mainly) on such problems (and I do not recommend it to my associates). A physicist can, and actually does, find something important and interesting to him or her personally in many problems of varying general significance, but this does not contradict the selection of a number of problems significant for the development of physics as a whole.

Now, a few remarks on the "human factor" in the immediate sense of the word.

Natural sciences study nature, the numerous natural objects and processes, and the laws governing them. Hence, science, for instance physics, is not at all affected by the cognoscitive subjects (in a philosopher's parlance). But it is precisely these subjects, men and women, who work in science and there are now millions of them. Some types of research have industrial and economic links, involve considerable expenditure and so on. Therefore, the development of science is closely linked to politics, economics, technology, sociology and psychology and has numerous humanitarian implications. These interrelationships are often very complex, hard to understand and to analyse. This is why they receive relatively little attention, at least in scientific publications. Incidentally, the literary style in science has been greatly affected by the desire (which is largely justifiable and natural) to get rid of everything that is not essential to the matter in hand, to shed off all extraneous relationships. A typical illustration of this process, although not the most important one, is the elimination of the personal pronoun I from scientific literature. For instance, I just cannot write "I think" in a purely scientific paper, and it has sometimes been an effort to do so in this book, which is, after all, a popular account of my personal views; and to write "in our opinion" or "one assumes" could even seem silly.

But the significance of the human factor is not diminished if we try to ignore it and concentrate on science itself.

A hidden tape-recorder placed in a laboratory would reveal, probably, that scientists or students talk about purely scientific matters no more than half of the time.

They also discuss what research fields are more promising, what to choose for work, what is now most important and interesting or attractive (or even profitable, convenient and so on). All these questions are quite natural.

I came to the decision to write this book with only one aim in mind. I thought that there was so much of interest in various fields of physics and astrophysics, and yet many budding physicists or students were not aware of it, and could not find out for themselves. So I decided to do something constructive in this way, to describe, at least briefly, some current problems of physics and astrophysics. Then complications arose. It was not clear how to select the problems, what standards to use and, finally, to whom the book should be addressed. These difficult questions have been discussed in the Prefaces and Introduction to the book and even now as I end this edition, which is a product of several revisions, I still cannot furnish a clear and definite answer to them. This is the reason for the numerous reservations in the book as, above all, I tried to avoid misunderstandings or wrong conclusions.

Perhaps the most erroneous conclusion would be to suspect that the author tries to lecture and to impose his opinions as to what is "important and interesting" and what is not. On the contrary, I think that such a sensitive issue is bound to produce controversy and differences of opinion. A more or less general consensus contributing to the development of science can be reached only by constructive debate, where arguments and counterarguments are freely and collectively discussed and compared, unclear and controversial issues are identified, and the truth is, if not arrived at, then at least approximated. It should be added, though, that there are all kinds of debate, and some people tend to regard their scientific opponents as enemies who must be insulted, humiliated and, if possible, destroyed (unfortunately, such attitudes are not unusual). What I call for is a debate on the development of science, taking place in an atmosphere of goodwill and tolerance and free of unhealthy excitement. Particularly, I would like to urge my colleagues, physicists and astrophysicists, to speak and write more often on the general problems of the development of science. Apart from other benefits, this will make it possible for a wide-reading public to get acquainted with different views and thus to make their own well-founded conclusions.

References

So many problems are discussed in the book that it would be unrealistic to attempt to compile a complete list of references. Therefore I refer primarily to review papers and to some original papers I had at hand when preparing the manuscript. This is the only reason for a comparatively high proportion of my own papers in the list of references.

1. (a) Ginzburg, V. L., *Theoretical Physics and Astrophysics*, Pergamon Press 1979). (b) Ginzburg, V. L., Tsytovich, V. L., *Physics Reports*, **49**, No 1, 1 (1979); *Physics Letters*, **79/A**, 16 (1980).
2. Dyson, F., *Physics Today*, **23**, No. 9, 23 (1970).
3. (a) *Nature of Matter*. Purposes of High Energy Physics, *Usp. Fiz. Nauk*, **86**, 591 (1965); Markov, M. A., *Usp. Fiz. Nauk*, **111**, 719 (1973); (b) Markov, M. A., *O prirode materii*, Moscow, Nauka (1976).
4. Anderson, P., *New Scientist and Sci. J.*, **51**, 510 (1971).
5. Phillips, J. C., *Comm. Solid State Phys.*, **4**, 91 (1972).
6. Kadomtsev, B. B., Strelkov, V. C., *Vest. Akad. Nauk SSSR*, No. 11, 67 (1980); Hagler, M. O., and Kristiansen, M., *An Introduction to Controlled Thermonuclear Fusion*, Lexington Books, D. C. Heath and Co., (1977); Stickley, C. M., *Physics Today*, **31**, No. 5, 50 (1978); Arnold, R. C., *Nature*, **276**, 19 (1978); *Physics Today*, **32**, No. 5 (1979).
7. Ginzburg, V. L., *Usp. Fiz. Nauk* **95**, 91 (1968); **101**, 185 (1970); **118**, 315 (1976) English translations: *Sov. Phys. — Uspekhi (USA)*.
8. *High-Temperature Superconductivity*. Eds. Ginzburg, V. L., and Kirzhnits, D. A., Plenum Press, (1982).
9. (a) Phillips, J. C., *Phys. Rev. Lett.*, **29**, 1551 (1972); (b) Cohen, M. L., Anderson, P. W., *Superconductivity of d- and f-Band Metals:* AIP Conference Rochester, N. Y. (1971); 17 (1972).
10. Kirzhnits, D. A., *Usp. Fiz. Nauk*, **119**, 357 (1976); Dolgov, O. V., Maksimov, E. G., *Pis'ma Zh. Eksp. Teor. Fiz.*, **28**, 3 (1978).
11. Brandt, N. B., Kuvshinnikov, S. V., Rusakov, A. P., Semenov, M. V., *Pis'ma Zh. Eksp. Teor. Fiz.*, **27**, 37 (1978).
12. Chu, C. W., Rusakov, A. P., et al., *Phys. Rev.*, **B18**, 2116 (1978).
13. Volkov, B. A., Ginzburg, V. L., Kopayev, Y. V., *Pis'ma Zh. Eksp. Teor. Fiz.*, **27**, 221 (1978); Volkov, B. A., Kopayev, Y. V., Tugushev, V. V., *Pis'ma Zh. Eksp. Teor. Fiz.*, **27**, 615 (1978); **30**, 317 (1979); Ginzburg, V. L., *Pis'ma Zh. Eksp. Teor. Fiz.*, **30**, 345 (1979). Ginzburg V. L. et al., *Solid State Commun.* **50,** 339 (1984).
14. Stepanov, A. V., *Fiz. Tverdogo Tela*, **1**, 671 (1959).
15. Schneider, T., *Helv. Phys. Acta*, **42**, 957 (1969).
16. Brovman, E. G., Kagan, Y. M., Kholas, A., *Zh. Eksp. Teor. Fiz.*, **61**, 2429 (1971); **62**, 1492 (1972); see also *Zh. Eksp. Teor. Fiz.*, **73**, 967 (1977).
17. Ginzburg, V. L., *Usp. Fiz. Nauk*, **97**, 601 (1969).
18. (a) Salpeter, E. E., *Phys. Rev. Lett.*, **28**, 560 (1972); Chapline, G. F., *Phys. Rev.*, **B6**, 2067 (1972); (b) Harris, F. E., Delhalle, J., *Phys. Rev. Lett.*, **39**, 1340 (1977); Avilov, V. V., Iordansky, S. V., *Fiz. Tverdogo Tela*, **19**, 3516 (1977); Mon, K. K., Chester, G. V., and Aschroft, N. W., *Phys. Rev.* **B21**, 2461 (1980).
19. (a) Vereshchagin, L. F., Yakovlev, E. N., Timofeev, Y. A., *Usp. Fiz. Nauk*, **117**, 183 (1975); *Pis'ma Zh. Eksp. Teor. Fiz.*, **21**, 190 (1975); (b) Hawke, R. S., et al., *Phys. Rev. Lett.*, **41**, 994 (1978);

Grigor'yev, F. V., Kormer, S. B., et al., *Zh. Eksp. Teor. Fiz.*, **75**, 1683 (1978); Stishev, S. M., *Usp. Fiz. Nauk*, **127**, 719 (1979).

20. Keldysh, L. V., *Usp. Fiz. Nauk*, **100**, 514 (1970).
21. Ginzburg, V. L., *Usp. Fiz. Nauk*, **108**, 749 (1972); **113**, 335 (1974); Ginzburg, V. L, Kelle, V. V., *Pis'ma Zh. Eksp. Teor. Fiz.*, **17**, 428 (1973); Klyuchnik, A. V., Lozovik, Y. E., *Fiz. Tverdogo Tela*, **20**, 625 (1978).
22. Thomas, G. A., *Sci. Amer.*, **234**, No. 6, 28 (1976); Andryushin, E. A., Silin, A. P., *Low Temp. Phys.*, **3**, 1365 (1977); Rice, T. M. et al., *Solid State Phys.*, **32**, 1, (1977); Rice, T. M., *Contemp. Phys.*, **20**, 241 (1979).
23. Andryushin, E. A., Keldysh, L. V., Silin, A. P., *Zh. Eksp. Teor. Fiz.*, **73**, 1163 (1977); Lerner, I. V., Lozovik, Y. E., *Phys. Lett.*, **A 64**, 48 (1978); Kuramoto, Y., Morimoto, M., *Phys. Soc. Japan*, **44**, 1759 (1978); Andryushin, E. A., et al., *Zh. Eksp. Teor. Fiz.*, **79**, (1980).
24. Anderson, P. W., Edwards, S. F., *J. Phys.*: F, **5**, 965, (1975); Suzuki, M., *Progr. Theor. Phys.*, **58**, 1151 (1977); Kinzel, W., Fischer, K. H., *J. Phys.*: F, **7**, 2163 (1977); Morandi, G., Corbelli, G., *Lett. Nuovo Cimento*, **24**, 273 (1979); Volovik, G. E., Dzyaloshinsky, I. E., *Zh. Eksp. Teor. Fiz.*, **75**, 1102 (1978); Bray, A. J. and Moore, M. A., *J. Phys.*, C: *Solid State Phys.*, **13**, 419 (1980); Blondin, A., Gabay, M., Garel, T., *Solid State Phys.*, **13**, 403 (1980); Parisi, G., *J. Phys.*, *A*: *Math. Gen.*, **13**, 1101 (1980).
25. Mott, N. F., *Contemp. Phys.* **18**, 225 (1977); Chaudhari, P., Turnall, D., *Science*, **199**, 11 (1978); Gilman, J. J., *Science*, **208**, 856 (1980).
26. Landau, L. D., Lifshitz, E. M., *Statistical Physics*, Oxford, Pergamon Press (1969). Ch 14.
27. Reatto, L., *J. Low Temp. Phys.*, **2**, 353 (1970); Berezinsky, V. L., *Zh. Eksp. Teor. Fiz.*, **59**, 907 (1970); **61**, 1144 (1971); Griffiths, R. B., *Phys. Rev. Lett.*, **23**, 17 (1969); McCoy, B. M., *Phys. Rev Lett.*, **23**, 383 (1969); Kosterlitz, J. M.., Thouless, D. J., *J. Phys.*: C, **6**, 1181 (1973); Lozovik, Y. E and Yudson, V. I., *Physica*, **93A**, 493 (1978).
28. Stanley, H., *Introduction to Phase Transitions and Critical Phenomena*, Oxford, Clarendon Press 1971; Patashinsky, A. Z., Pokrovsky, V. L., *Fluctuation Theory of Phase Transitions*, Moscow, Nauka, 1975 (in Russian); Levelt Sengers, A. N., Hocken, R., Sengers, J. V., *Physics Today*, **30**, No. 12, 42 (1977); Nelson, D. R., *Nature*, **269**, 379 (1977).
29. Ginzburg, V. L., *Fiz. Tverdogo Tela*, **2**, 2031 (1960); Levanyuk, A. P., Sobyanin, A. A., *Pis'ma Zh. Eksp. Teor. Fiz.*, **11**, 540, (1970); Vaks, V. G., Larkin, A. I., Pikin, S. A., *Zh. Eksp. Teor. Fiz.*, **51**, 361, (1966); **56**, 2087, (1969); Bausch, B., *Z. Phys.*, **254**, 81 (1972); **258**, 423 (1973).
30. Levelt Sengers, J. M. H., *Physica*, **82A**, 319 (1976).
31. Ginzburg, V. L., Sobyanin, A. A., *Usp. Fiz. Nauk*, **120**, 153, 733, (1976); Ginzburg, V. L., Sobyanin, A. A., *Phys. Lett.*, **A69**, 417, (1979). *T., Low Temp Phys.* **49**, 507 (1982).
32. Yakovlev, I. A., Velichkina, T. S., *Usp. Fiz. Nauk*, **63**, 411, (1957).
33. Ginzburg, V. L., *Usp. Fiz. Nauk*, **77**, 621 (1962); Levanyuk, A. P., Sobyanin, A. A., *Zh. Eksp. Teor. Fiz.*, **53**, 1024, (1967).
34. Shapiro, S. M., Cummins, H. Z., *Phys. Rev. Lett.*, **21**, 1578 (1968); Fritz, I. J., Cummins, H. Z., *Phys. Rev. Lett.*, **28**, 96 (1972).
35. Ginzburg, V. L., Levanyuk, A. P., Sobyanin, A. A., *Usp. Fiz. Nauk*, **130**, 615 (1980); *Phys. Rep.*, **57**, No. 3, 151, (1980).
36. Fabelinsky, I. L., *Molecular Light Scattering*, New York, Plenum Press, 1968; Gorelik, V. S., Sushchinsky, M. M., *Usp. Fiz. Nauk*, **98**, 237, (1969).
37. Mermin, N. D., *Phys. Rev.*, **A9**, 868 (1974); *Quantum Fluids and Solids*, Eds. S. B. Trickey, E. D. Adams, J. W. Duffy, Plenum Press, New York, London, 1977; Wheatley, J., *Physics Today*, **29**, No. 2, 32 (1976).

38. (a) Ginzburg, V. L., Sobyanin, A. A., *Pis'ma Zh. Eksp. Teor. Fiz.*, **15**, 343 (1972); Akulichev, V. A., Bulanov, V. A., *Zh. Eksp. Teor. Fiz.*, **65**, 668 (1973); *Akustichesky zhurnal*, **20**, 817 (1974); (b) Andreev, A. F., Lifshitz, E. M., *Zh. Eksp. Teor. Fiz.*, **56**, 2057 (1969); Greeywall, D. S., *Phys. Rev.*, **B16**, 1291 (1977).
39. (a) Kadomtsev, B. B., Kudryavtsev, V. S., *Pis'ma Zh. Eksp. Teor. Fiz.*, **13**, 15, 61 (1971); (b) Ginzburg, V. L., Usov, V. V., *Pis'ma Zh. Eksp. Teor. Fiz.*, **15**, 280 (1972); (c) Hillbrandt, W., Müller. E., *Astrophys. J.*, **207**, 589 (1976).
40. Goldansky, V. I., Kagan, Y. M., *Zh. Eksp. Teor. Fiz.*, **64**, 90 (1973); *Usp. Fiz. Nauk*, **110**, 445 (1973); Letokhov, V. S., *Zh. Eksp. Teor. Fiz.*, **64**, 1555 (1973); Il'insky, Y. A., Khokhlov, R. V., *Usp. Fiz. Nauk*, **110**, 449 (1973); Il'insky, A. A., Priroda, No. 9, 49 (1978); Chaplin, G., Wood, L., *Phys. Today*, **28**, (6), 41 (June 1975); Karyagin, S. V., *Zh. Eksp. Teor. Fiz.*, **79**, 731 (1980).
41. (a) Uman, M., *Lightning*, New York, McCraw-Hill, 1969; (b) Singer, S., *The Nature of Ball Lightning*, New York, Plenum Press, 1971; (c) Wooding, E. R., *Nature*, **239**, 394 (1972); Crawford, J. E., *Nature*, **239**, 395 (1972).
42. Stakhanov, I. P., *Pis'ma Zh. Eksp Teor. Fiz.*, **18**, 193 (1973); *Zh. Tech. Fiz.*, **44**, 1373 (1974); **46**, 82 (1976); Smirnov, B. M., *Usp. Fiz. Nauk*, **116**, 731 (1975); Stakhanov, I. P., *Physical Nature of Ball Lightning*, Moscow, Atomizdat, 1979 (in Russian); Barry, J. D., *Ball Lightning and Bead Lightning*, Plenum Press, N. Y. and London, 1980.
43. de Gennes, P. G., *The Physics of Liquid Crystals*, Oxford, Clarendon Press, 1974; Pikin, S. A., Indenbom, V. L., *Usp. Fiz. Nauk*, **125**, 251 (1978).
44. (a) Duke, C. B., Park, R. L., *Phys. Today*, **25**, (8) 23 (1972); Somorgiai, G. A., *Science*, **201**, 489 (1978); Czanderna, A. W., *Methods of Surface Analysis*, Amsterdam, 1975; (b) Agranovich, V. B., Ginzburg, V. L., *Crystal Optics with Spatial Dispersion and the Theory of Excitons*, Berlin, Springer, in press.
45. Flerov, G. N., Druin, V. A., Pleve, A. A., *Usp. Fiz. Nauk*, **100**, 45 (1970); Flerov, G. N., *Priroda*, No. 9, 56 (1972); Flerov, G. N., et al., *Yadernaya Fizika*, **26**, 449 (1977).
46. Seaborg, G., Bloom, D., *Sci. Amer.*, **220**, No. 4, 56, (1968); Ghiorso, A., et al., *Phys. Rev. Lett.*, **22**, 1317 (1969); **24**, 1498 (1970); Price, P. B., Fleischer, R. L., Woods, R. T., *Phys. Rev.*, **61**, 1819 (1970); Anders, E., Larimer, J. W., *Science*, **175**, 981 (1972): Nix, J. R., *Usp. Fiz. Nauk*, **110**, 405 (1973); Zhdanov, G. B., *Usp. Fiz. Nauk*, **111**, 109 (1973); Seaborg, G. T., *Science*, **203**, 711 (1979); Hermann, G., *Nature*, **280**, 543 (1979).
47. Polikanov, S. M., *Usp. Fiz. Nauk*, **107**, 685 (1972).
48. Shapiro, I. S., *Usp. Fiz. Nauk*, **125**, 577 (1978).
49. Migdal, A. B., *Usp. Fiz. Nauk*, **123**, 369 (1977);
Goldhaber, A. S., *Nature*, **275**, 115 (1978); Ericson, M., Delorme, J., *Phys. Lett.*, **76A**, 182 (1978).
50. Müller, E., *Usp. Fiz. Nauk*, **92**, 293 (1967).
51. *Physics Today*, **23**, No. 8, 41 (1970).
52. Einstein, A., "Notes on origin of the general theory of relativity", in *Ideas and Opinions*, London, Redman (1956), p. 285 (translated from the German).
53. Weisskopf, V. F., *Sci. Amer.*, **218**, No. 5, 15 (1968); **104**, 131 (1971).
54. Review of particle properties, *Rev. Mod. Phys.*, **48**, Pt. 2 (1976).
55. Richter, B., From the Psi to Charm—The Experiments of 1975 and 1976. Nobel Lecture, December 1976, in *Les Prix Nobel en 1976*, Stockholm, 1977, pp. 42–74; Ting, S. C. C., *Les Prix Nobel en 1976*, Stockholm, 1977, pp. 76–106; Lederman, L. M., *Sci. Amer.*, **239**, No. 4, 60–80 (October 1978); Schwitters, R. F., *Sci. Amer.*, **237**, No. 4, 56 (1977).
56. (a) Glashow, Sheldon Lee, *Sci. Amer.*, **233**, (4), 38 (October 1975); (b) Nambu, Y., *Sci. Amer.*, **235** (5), 48–60 (October 1976); Mulvey, J., *Nature*, **278**, 403 (1979); Marciano, W., and Pagels, H., *Nature*, **279**, 479 (1979); Zichichi, A., *Rev. Nuovo Cim.*, **2**, No. 14, 1, (1979).

57. (a) Kendall, G., Panofsky, W., *Sci. Amer.*, **224**, No 6, 60 (1976); (b) Drell, S., *Comm. Nucl. Part. Phys.*, **4**, No. 4, 147 (1970); (c) Litke, A. M., Wilson, R., *Sci. Amer.*, **229**, No. 4., 104 (1973).
58. Drell, S. D., *Physics Today*, **31**, No. 6, 23 (1978).
59. Salam, A., *Proc. Roy. Soc.*, **A355**, 515 (1977).
60. Heisenberg, W., *Phys. Today*, **29**, (3), 32 (March 1976).
61. Sachs, R. G., *Science*, **176**, 587 (1972).
62. Heisenberg, W., *Introduction to the Unified Field Theory of Elementary Particles*, USA, 1966.
63. Kirzhnits, D. A., *Usp. Fiz. Nauk*, **125**, 169 (1978); see also Linde, A. D., *Rep. Progr. Phys.*, **42**, 389 (1979).
64. Ginzburg, V. L., Man'ko, V. I., *Sov. Journ. Part. Nucl.*, **7**, 1 (1976).
65. Perl, M. L., Kirk, W. T., *Sci. Amer.*, **238**, No. 3, 50 (1978); Perl, M. L., *Nature*, **275**, 273 (1978); Azimov, Ya. I., Khoze, V. A., *Usp. Fiz. Nauk*, **132**, 379 (1980).
66. Riemann, B., "On the hypotheses that lie at the basis of geometry," *Nature*, **183**, 14 (1973) (translated from the German).
67. Einstein, A., "Geometry and experience" in *Ideas and Opinions*, London, Redman 232 (1956).
68. Snyder, H., *Phys. Rev.*, **71**, 38 (1974); Tamm, I. E., *Vest. Akad. Nauk*, No. 9, 22 (1968); *Problemy Teoreticheskoy Fiziki:* Sbornik pamyati I. E. Tamma, Moscow, Nauka (1972).
69. Feinberg, E. L., *Usp. Fiz. Nauk*, **86**, 733 (1965); Amaldi, U., *Sci. Amer.*, **229**, No. 5, 36 (1973).
70. Amaldi, U., *Usp. Fiz. Nauk*, **124**, 651 (1978).
71. Feinberg, E. L., *Usp. Fiz. Nauk*, **104**, 539 (1971); Usp. Fiz. Nauk, **132**, 255 (1980).
72. Bugaev, E. V., Kotov, Y. D., Rosental, I. L., "Cosmic Muons and Neutrinos", Moscow, Atomizdat, 1970; Ermolov, P. V., Mukhin, A. I., *Usp. Fiz. Nauk*, **124**, 385 (1978).
73. Cline, D. C., Mann, A. K., Rubbia, C., *Sci. Amer.*, **234** (1), 44 (January 1976).
74. "The Problem of CP Violation", *Usp. Fiz. Nauk*, **95**, 401 (1968); Swetman, T. P., *Amer. J. Phys.*, **39**, 1320 (1971); Sachs, R. G., *Science*, **176**, 587 (1972); Barish, B. C., *Sci. Amer.*, **229**, No. 2, 30 (1973); Nanopoulos, D. V., Yildiz, A., and Cox, P. H., *Annals Phys.*, **127**, 126 (1980).
75. Weinberg, S., *Sci. Amer.*, **231** (1), 50 (July 1974); Clain, D. B., Mann, A. K., Rubbia, C., *Sci. Amer.*, **231** (6), 108 (December 1974).
76. Freedman, D. Z., van Nieuwenhuizen, P., *Sci. Amer.*, **238**, No. 2, 126 (1978).
77. Iliopoulus, J., *An Introduction to Gauge Theories*, CERN, Preprint 76–11, Geneva, June 1976, p. 42; Slavnov, A. A., *Usp. Fiz. Nauk*, **124**, 487 (1978); Abers, E. S., Lee, B. W., *Phys. Reports (Section C of Phys. Lett.)*, **9**, 1 (1973).
78. *Nature*, **274**, 11 (1978); Feinberg, G., *Nature*, **271**, 509 (1978).
79. Weinberg, S., *Physics Today*, **30**, No. 4, 42 (1977).
80. *Nature*, **267**, 9 (1977); Mulvey, J., *Nature*, **278**, 403 (1979); Wilson, R. R., *Sci. Amer.*, **242**, No. 1, 26 (1980).
81. McKay, D. W., *Phys. Rev.*, **D16**, 2861 (1977); Mohapatra, R. N., Sidhu, D. P., *Phys. Rev.*, **D17**, 876 (1978).
82. Nikishev, A. A., Ritus, V. I., *Quantum Electrodynamics of the Phenomena in High-Intensity Fields*, Moscow, Trudy FIAN, **111**, 1979.
83. Zeldovich, Ya. B., Popov, V. S., *Usp. Fiz. Nauk*, **105**, 403 (1971).
84. Lee, H. C., Khanna, F. C., *Canad. J. Phys.*, **56**, 149 (1978); Klein, A., Rafelsky, J., *Z. Physik*, **A284**, 71 (1978).
85. Ginzburg, V. L., *Usp. Phys. Nauk*, **80**, 207 (1963).
86. (a) Rudenko, V. N., *Usp. Fiz. Nauk*, **126**, 361 (1978); (b) Konopleva, N. P., *Usp. Fiz. Nauk*, **123**, 577 (1977).
87. (a) Ginzburg, V. L., *Usp. Fiz. Nauk*, **128**, 435 (1979); (b) Ginzburg, V. L., *On the Relativity Theory*, Moscow, Nauka, 1979.

88. Landau, L. D., Lifshitz, E. M., *The Classical Theory of Fields*, Oxford, Pergamon Press (1971), p. 386.
89. Einstein, A., *Sitzungsher. Preuss. Akad. Wiss.*, 1916, 688.
90. Press, W. H., Thorne, K. S., *Ann. Rev. Astron. and Astrophys.*, **10**, 335 (1972); "Gravitational-wave Astronomy", Preprint OAP-**273**, Caltech (1972); Grishchuk, L. P., *Usp. Fiz. Nauk*, **121**, 629 (1977); Tyson, J. A., Giffard, R. P., *Ann. Rev. Astron. Astrophys.*, **16**, 521 (1978).
91. Braginsky, V. B., Manukin, A. B., *Measurement of Low Forces in Physical Experiments*, Moscow, Nauka, 1974 (in Russian); Braginsky, V. B., Rudenko, V. N., *Phys. Rep.*, **46**, 165 (1978); Braginsky, V. B., Vorontsov, Y. I., Khalili, F. Ya., *Pis'ma Zh. Eksp. Teor. Fiz.*, **27**, 296 (1978); Thorne, K. S., in *Theoretical Principles in Astrophysics and Relativity*, Ed. Lebovich, N. R., et al., Chicago, **149**, 1978; Hellings, R. W., *Phys. Rev.*, **D17**, 3158 (1978).
92. Weber, J., *Phys. Rev. Lett.*, **22**, 1320 (1969); **25**, 180 (1970); *Nature*, **240**, 28 (1972).
93. (a) Friedman, A. A., *Zeitschrift für Physik*, **11**, **377** (1922); **21**, 326 (1924); (b) Zeldovich, Ya. B., Novikov, I. D. *The Structure and Evolution of Universe*, Chicago Univ. Press, in press; (c) Weinberg, S., *Gravitation and Cosmology*, New York, 1972; (d) Sciama, D., *Modern Cosmology*, Cambridge, 1971; (e) Peebles, P., *Physical Cosmology*, Princeton, 1971; (f) Longair, M. S., *Q. J. Roy. Astr. Soc.*, **17**, 422 (1976).
94. Einstein, A., *Sitzungsher. Preuss. Akad. Wiss.*, **1**, 142–152 (1917).
95. Markov, M. A., *Usp. Fiz. Nauk*, **111**, 3 (1973).
96. Wheeler, J., *Einstein's Vision*, New York, Springer-Verlag, 1968 (in German); Ginzburg, V. L., Kirzhnits, D. A., Lyubushin, A. A., *Zh. Eksp. Teor. Fiz.*, **60**, 451 (1971); "Gravitation", Kiev, Naukova dumka, (1972), p. 40; Parker, L., *Phys. Rev. Lett.*, **28**, 705 (1972); *Phys. Rev.*, **D7**, 2357 (1973).
97. Hu, B. L., Parker, L., *Phys. Rev.*, **D17**, 933 (1978); Ford, L. H., Parker, L., *Phys. Rev.* **D17**, 1485 (1978); Vereshkov, G. M., et al., *Zh. Eksp. Teor. Fiz.*, **73**, 1985 (1977); Zeldovich, Ya. B., *Usp. Fiz. Nauk*, **123**, 487 (1977); Gibbons, G. W., Hawking, S. M., *Phys. Rev.*, **D15**, 2738 (1977).
98. Penrose, R., in *Theoretical Principles in Astrophysics and Relativity*, Ed. Lebovich, N. R., et al., Chicago, 1978, p. 217.
99. Canuto, V., Hsien, S. H., Adams, P. J., *Phys. Rev. Lett.*, **39**, 429 (1977); *Astrophys. J.*, **224**, 302 (1978).
100. Steigman, G., *Ann. Rev. Astron. Astrophys.*, **14**, 339 (1976); Stecker, F. W., *Nature*, **273**, 493 (1978).
101. Jeans, J. H., *Astronomy and Cosmogony*, Cambridge; Cambridge Univ. Press, 1928, p. 352.
102. (a) *Origin and Evolution of Galaxies and Stars*, Ed. Pikelner, S. B., Moscow, Nauka, 1976 (in Russian); (b) Ozernoy, L. M., *Origin and Life of Galaxies*, Moscow, Znanie, 1978. (c) Zonn, W., *Galaktyki i Kwazary*, Warsawa, 1975; (d) Oort, J. H., *Ann. Rev., Astron. Astrophys.*, **15**, 295 (1977).
103. (a) Ambartsumyan, V. A., *Usp. Fiz. Nauk*, **96**, 3 (1968); *The Structure and Evolution of Galaxies*, in Proc. 13th Solvay Conf. on Physics, Intersci. Publ., 1965, p. 1; (b) Bahcall, J. N., Joss, P. C., *Comm. Astrophys. a. Space Phys.*, **4**, 95 (1972).
104. Ginzburg, V. I., Ozernoy, L. M., *Astrophys. a. Space Sci.*, **48**, 401 (1977); Ozernoy, L. M., *Usp. Fiz. Nauk*, **120**, 309 (1976); Bailey, M. E., Clube, S. V. M., *Nature*, **275**, 278 (1978); *Active Galactic Nuclei*, Eds. C. Hazard and S. Mitton, Cambridge University Press, Cambridge, 1979.
105. (a) Ostriker, J. P., *Proc. Nat. Acad. Sci.* USA, **74**, 1767 (1977); (b) *The Large Scale Structure of the Universe*, Ed., Longeir, M. S., Einasto, J., in IAU Symposium, **79**, 1978; Spinard, H., et al., *Astrophys. J.*, **225**, 56 (1978).
106. Cowsik, R., McClelland, J., *Astrophys. J.*, **180**, 7 (1973);
107. Ginzburg, V. L., *D. J. Roy. Astr. Soc.*, **16**, 265, (1975).
108. Baade, W., Zwicky, F., *Proc. Nat. Acad. Sci.* USA, **20**, 259 (1934).

109. (a) Zeldovich, Ya. B., Novikov, I. D., *Relativistic Astrophysics*, **1**, *Stars and Relativity*, Chicago Univ. Press, 1971; (b) Baym, G., Pethick, C. J., *Ann. Rev. Nucl. Sci.*, **25**, 27 (1975); *Ann. Rev. Astron. Astrophys.*, **17**, 415 (1979); Lamb, F. K., *Ann. New York Acad. Sci.*, **302**, 482 (1977).
110. Hewish, E., *Sci. Amer.* **219**, No. 4 (1968), p. 25; Ginzburg, V. L., *Usp. Fiz. Nauk*, **103**, 393 (1971); D. ter Haar, *Contemp. Phys.*, **16**, 243 (1975); Ginzburg, V. L., Zheleznyakov, V. V., *Ann. Rev. Astron. Astrophys.* **13**, 511 (1975); Manchester, R. N., Taylor, J. H., *Pulsars*, San Francisco: Freeman and Co., 1977.
111. Usov, V. V., *Galactical and Extragalactical Astronomy: High-Energy Astrophysics*, Moscow, VINITI, 1977.
112. Kirzhnits, D. A., *Usp. Fiz. Nauk*, **104**, 489 (1971); Canuto, V., *Ann Rev. Astron. Astrophys.*, **13**, 335 (1975); (see also 17); Sauls, J. A., Serene, J. W., *Phys. Rev.* **D17**, 1524 (1978); Takatsuka, J., et al., *Progr. Theor. Phys.*, **59**, 1933 (1978); Chela-Flores, J., *Phys. Rev.* **D18**, 2632 (1978).
113. Treumper, J., et al,. *Astrophys. J. (Letters)*, **219**, L 105 (1978); Coe, M. J., et al., *Nature*, **268**, 508 (1977).
114. Gerstein, S. S., et al,. *Pis'ma Zh. Eksp. Teor. Fiz*, **26**, 189 (1977); Ivanova, L. N., Imshennik, V. S., Chechetkin, V. M., *Astron. Zh.* **54**, 1009 (1977); Barkt, Z., *Rev. Astrophys.*, **13**, 45 (1975); Arnett, W. D., *Ann. Rev. Astron. Astrophys.*, **11**, 73 (1973).
115. Gursky, G., van der Heuvel, E., *Sci. Amer.*, **232**, No. 3, 24 (1975); Rappaport, S., Joss, P. C., *Nature*, **266**, 123 (1977).
116. Freedman, B., McLerran, L., *Phys. Rev.*, **D17**, 1109 (1978); Fechner, W. P., Joss, P. C., *Nature*, **274**, 347 (1978).
117. Thorne, K. S., *Sci. Amer.*, **231** (6), 32 (1974) Penrose, R., *Sci. Amer.*, **226**, (5), 38 (1972); Frolov, V. P., *Usp. Fiz. Nauk*, **118**, 473 (1976); De Witt, B. S., *Physics Reports*, **19C**, 295 (1975).
118. Hawking, S. W., *Nature*, **248**, 30 (1974); *Comm., Math. Phys.*, **43**, 199 (1975); *Phys. Rev.*, **D13**, 191 (1976).
119. Ginzburg, V. L., *Usp. Fiz. Nauk*, **59**, 11 (1956).
120. Laitman, A. P., Syunyaev, R. A., et al., *Usp. Fiz. Nauk*, **126**, 515 (1978); see also Samimi, J., et al., *Nature*, **278** 434 (1979); Evans, W. D., et al., *Nature*, **278**, 434 (1979).
121. Dokuchaev, V. I., Ozernoy, L. M., *Pis'ma Astron. Zh.*, **3**, 391 (1977); *Astron. Zh.*, **55**, 27 (1978).
122. Sargent, W. L. W., Young, P., et al., *Astrophys. J.*, **221**, 721, 731 (1977); see also *Nature*, **274**, 419 (1978).
123. De Felice, F., *Nature*, **273**, 429 (1978).
124. Oppenheimer, J. R., Snyder, H., *Phys. Rev.*, **56**, 455 (1939).
125. Lohiya, D., Panchapakesan, *Lett. Nuovo Cimento*, **21**, 81, (1978); Lake, K., Roeder, R. C., *Nature*, **273**, 449 (1978).
126. Rees, M. J., *Nature*, **226**, 333 (1977); Blandford, R. D., *Mon. Not. RAS*, **181**, 489 (1977); Porter, N. A., Weeks, T. C., *Mon. Not. RAS*, **183**, 205 (1978).
127. Ginzburg, V. L., *Pis'ma Zh. Eksp. Teor. Fiz.*, **22**, 514 (1975); Ginzburg, V. L., Frolov, V. P., *Pis'ma Astron. Zh.*, **2**, 474 (1976).
128. (a) Weeks, T., *High-Energy Astrophysics*, London, 1969; (b) Hayakawa, S., *Cosmic Ray Physics. Nuclear and Astrophysical Aspects*, New York, 1969; (c) Ozernoy, L. M., Prilutsky, O. F., Rozental', I. L., *High-Energy Astrophysics*, Moscow, Atomizdat, 1973 (in Russian).
129. (a) Ginzburg, V. L., Ptuskin, V. S., *Usp. Phys. Nauk*, **117**, 585 (1976); Ginzburg, V. L., *Usp. Phys. Nauk*, **124**, 307 (1978); (b) Ginzburg, V. L., Dorman, I. V., *Priroda*, No. 4, 10 (1978).
130. Kaplan, S. A., Tzytovich, V. N., *Plasma Astrophysics*, Oxford, Pergamon Press (1973); Ginzburg, V. L., Ptuskin, V. S., Tzytovich, V. N., *Astrophys. a. Space Sci.*, **21**, 13 (1973).

131. Dorman, L. I., *Experimental and Theoretical Fundamentals of Astrophysics of Cosmic Rays*, Moscow, Nauka, 1975 (in Russian).
132. Khristiansen, G. B., *High-Energy Cosmic Rays*, Moscow, MGU, 1974 (in Russian); Linsley, J., *Sci. Amer.*, **239**, No. 1., 48 (1978).
133. Levin, W. H. G., Joss, P. C., *Nature*, **270**, 211, 310 (1977); Forman, W., et al., *Astrophys. J.* (*Letters*), **225**, L 1 (1978); Helfand, D. J., et al., *Nature*, **283**, 337 (1980); *X-ray Astronomy (COSPAR).* Ed. Baity, W. A. and Peterson, L. E., Pergamon Press, 1979.
134 Morisson, P., *Nuovo Cimento*, **7**, 858 (1958).
135. Ginzburg, V. L., Syrovatsky, S. I., *Space Sci. Reviews*, **4**, 267 (1965); Fazio, G. G., *Ann. Rev. Astron. and Astrophys.*, **5**, 481 (1968); Gal'per, A. M., Kirillov-Ugryumov, V. G., Luchkov, B. I., *Usp. Fiz. Nauk*, **112**, 491 (1974); **128**, 313 (1979).
136. *Proc. 12th ESLAB Symposium of Astronomy: Recent Advances in Gamma-Ray Astronomy*, Frascati, 1977; *15th International Cosmic-Ray Conference:* Conference papers, Plovdiv, 1977; Pinkau, K., *Nature*, **277**, 17 (1979).
137. Lingenfelter, R. E., Ramaty, R., *Physics Today*, **31**, No. 3, 40 (1978); Rozental, I. L., Usov, V. V., Estulin, I. V., *Usp. Fiz. Nauk*, **127**, 135 (1979); Leventhal, M., et al., *Astrophys. J* (Letters), **225**, L 11 (1978).
138. (a) Klebensadel, R. W., Strong, I. B., Olsen, R. A., *Astrophys, J.* (Letters), **182**, L 85 (1973); (b) Cline, T. L., et al., *Astrophys. J.* (Letters), **185**, L1 (1973); (c) Wheaton, W. A., et al., *Astrophys., J.* (Letters), **185**, L 57 (1973).
139. Bahcall, J. N., *Sci. Amer.*, **211**, (1), 29 (1969); Wolfendale, A. W., *Neutrinos: Lecture to the British Association*, (Sept. 1970); Neutrino: Collected Papers, Moscow, Nauka, 1970 (in Russian).
140. *Neutrino 77: Proc. Intern. Conf. on Neutrino Physics and Neutrino Astrophysics*, Moscow, Nauka, 1978, **1**, Bahcall, J. N., et al., *Phys. Rev. Lett.*, **40**, 1351 (1978); Bahcall, J. N., *Rev. Mod. Phys.*, **50**, 881 (1978).
141. Bilen'ky, S. M., Pontokorvo, B. M., *Usp. Fiz. Nauk*, **123**, 181 (1977); Ulrich, R. K., *Science*, **190**, 619 (1975).
142. Berezinsky, V. S., Zatsepin, G. T., *Usp. Fiz. Nauk*, **122**, 3 (1977); Eichler, D., Schramm, D. N., *Nature*, **277**, 17 (1979); Miller, D. J., *Nature*, **280**, 191 (1979).
143. Ginzburg, V. L., *Modern Astrophysics*, Moscow, Nauka, 1970 (in Russian).
144. *Communications with extraterrestrial intelligence*, Ed. C. Sagan, MIT Press, Cambridge, Mass., 1973; Kuiper, W. B. H., Morris, M., *Science*, **196**, 166 (1977); Murray, B., Gulkis, S., Edelson, R. E., *Science*, **199**, 485 (1978); Bates, D. R., *Astrophys. a. Space Sci.*, **55**, 7 (1978); *Cosmic Search*, **1**, No. 1 (1979).
145. Kramarovsky, Ya. M., Chechev, V. P., *Usp. Fiz. Nauk*, **102**, 141 (1970); Dyson, F., *Sci. Amer.*, **225**, No. 9 51 (1971); Davies, P. C. W., *J. Phys.*: A **5**, 1296 (1972); Novello, M., Rotelli, P., *J Phys.*: A, **5**, 1488 (1972); Pagel, P. E. J., *Mon, Not. RAS*, **179**, 81 (1977); Blake, G. M., *Mon. Not. RAS*, **181**, 47 (1977); Ganguli, S. M., et al,. *Phys. Lett.*, **B74**, 130 (1978).
146. (a) Dirac, P. A. M., *Proc. Roy. Soc.*, **A165**, 199 (1938); **A333**, 419 (1973); (b) Dicke, R., in "*Gravitation and Relativity*", Eds. Hong Jee Ching, W. Hoffman, New York, Amsterdam (1964); (c) Braginsky, V. B., Ginzburg, V. L., *DAN SSSR*, **216**, 300 (1974); (d) Obregon, O. J., Pimentel, L. O., *General Relativity and Gravitation*, **9**, 585 (1978); Barrow, J. D., *Mon. Not. RAS*, **184**, 677 (1978); Blake, G. M., *Mon. Not. RAS*, **185**, 399 (1978); (e) Linde, A. D., *Pis'ma Zh. Eksp. Teor. Fiz.*, **30**, 479 (1979); (see also 96).
147. Ginzburg, V. L., *Priroda*, No. 6, 73 (1976).
148. Bromeley, D. A., *Physics Today*, **25**, No. 7, 23 (1972); *Astronomy and Astrophysics for the 1970s*, Rep. of the Astron. Survey Comm. Nat. Acad. Sci., Washington, (1972); *Sky and Telescope*, **44**, No. 6, 391 (1972).
149. Morse, Ph., *Physics Today*, **26**, No. 4, 23 (1973).
150. Brayshaw, D. D., *Phys. Rev. Lett.*, **42**, 1106 (1979).

151. Zweig, G., *Science*, **201**, 973 (1978).
152. Vessot, R. F. C., Levine, M. W., *General Relativity and Gravitation*, **10**, 181 (1979).
153. Taylor, J. H., Fowler, L. A., McCulloch, P. M., *Nature*, **277**, 437 (1979).
154. Islam, J. N., *Sky and Telescope*, **57**, 13 (1979); Barrow, J. D., Tipler, F. J., *Nature*, **276**, 453 (1978); Dyson, F. J., *Rev. Mod. Phys.*, **51**, **447** (1979).
155. Bartlett, A. A., *Amer. J. Phys.*, **46**, 876 (1978).
156. Leorned, J. R., Reines, F., Soni, A., *Phys. Rev. Lett.*, **43**, 907 (1979).
157. Jerome, D., Mazaud, A., Ribault, M. and Bechgard, K., *J. Phys. (Paris), Letts.*, **41**, 95 (1980); Andres, K., Wudl, F., McWhan, D. B., Thomas, G. A., Nalewajek, D. and Stevens, A. L., *Phys. Rev. Letts.*, **45**, 1449 (1980); Greene, R. L., and Engler, E. M., *Phys. Rev. Letts.*, **45**, 1587 (1980).
158. Lefkowitz, I., Manning, J. S., and Bloomfield, P. E., *Phys. Rev.*, *20 B*, 4506 (1979); *Phys. Rev. 23B*, 3022 (1981).
159. Gebale, T. E., and Chu, C. W., *Comments Solid State Phys.*, **9**, 115 (1979).
160. Brown, E., Homan, C. G., and MacCrone, R. K., *Phys. Rev. Letts.* **45**, 478 (1980).
161. Hardy, W. N., et al., *Phys. Rev. Letts.*, **45**, 453 (1980).
162. Walraven, J. T. M., Silvera, I. F., and Mattey, A. P. M., *Phys. Rev. Letts.*, **45**, 449 (1980); see also **45**, 915 (1980).
163. Pokrovsky, V. L., *Adv. Phys.*, **28**, 595 (1979).
164. Nelson, D. R., *Proc. Summer School in Statistical Mechanics*, Enschede, Netherlands, 1980.
165. *Proc. Intern. School on Condensed Matter Physics*, Varna, Bulgaria, 1980.
166. Lerner, I. V., and Lozovik, Y. E., *Zh. Eksp. Teor. Fiz.*, **78**, 1167 (1980).
167. Brantman, V. L., Ginzburg, N. S. and Petelin, M. I., *Optics Communs.*, **30**, 409 (1980); *Nature*, **285**, 15 (1980); *Science*, **204**, 394 (1979).
168. Tabor, D., *Surface Sci.*, **89**, 1 (1979).
169. Perelygin, B. P., and Stetsenko, S. G., *Pis'ma Zh. Eksp. Teor. Fiz.*, **32**, 622 (1980).
170. Blin-Stoyle, R. J., *Contemp. Phys.*, **20**, 377 (1979).
171. Pines, D., *Science*, **207**, 597 (1980); *J. de Phys.*, **41**, C2-111 (1980).
172. Alberico, W. M., et al., *Phys. Letts*, **92B**, 153 (1980); Vasak, D., et al., *Phys. Letts.*, **93B**, 243 (1980); Mohan, L. R. R. and Minich, R. W., *Phys. Letts*, **93B**, 467 (1980).
173. Cosslett, V. E., et al., *Nature*, **281**, 49 (1979).
174. Bartel, W., et al., *Phys. Letts.*, **89B**, 136 (1979); Barber, D. P., et al., *Phys. Rev. Letts.*, **44**, 1722 (1980).
175. Chanowitz, M. S., *Phys. Rev. Letts.*, **44**, 59 (1980).
176. Salom, A., *Rev. Mod. Phys.*, **52**, 525 (1980).
177. Okun, L. V., *Usp. Fiz. Nauk*, **133**, 3 (1981).
178. Glashow, Sh. L., *Rev. Mod. Phys.*, **52**, 539 (1980); Weinberg, S., *Rev. Mod. Phys.*, **52**, 515 (1980).
179. Bartel, W., et al., *Phys. Letts.*, **92B**, 206 (1980).
180. Nikol'sky, S. I., Feinberg, E. L., Avakyan, V. V., et al., *Usp. Fiz. Nauk*, **132**, 392, 394, 395 (1980); Feinberg, E. L., *Vestnik Akad. Nauk SSSR*, No. 1, 25 (1981).
181. Iliopoulos, J., *Contemp. Phys.*, **21**, 159 (1980); Hooft, G. t., *Scient, American*, **242**, No. 6, 90 (1980).
182. Cleine, D. and Rubbia, C., *Phys. Today*, **33**, No. 8 (1980).
183. Barkov, L. M., Zolotarev, M. S., and Khriplovich, I. B., *Usp. Fiz. Nauk*, **132**, 409 (1980).
184. Bogdanov, Y. B., Sobel'man, I. I., Sorokin, Y. N. and Struk, I. I., *Pis'ma Zh. Eksp. Teor. Fiz.*, **31**, 234, 556 (1980).
185. Berger, Ch., et al., *Phys. Letts.*, **86B**, 418 (1979); *Phys. Rev. Letts.*, **43**, 830 (1979); Azimov, Ya. I., Dokshitser, Y. L. and Khoze, B. A., *Usp. Fiz. Nauk.*, **132**, 443 (1980).
186. Veinstein, A. I., Zakharov, V. I., and Shifman, M. A., *Usp. Fiz. Nauk*, **131**, 537 (1980).

187. Georgi, H. and Glashow, S. L., *Phys. Today*, **33**, No. 9, 30 (1980) Georgi, H., *Sci Amer.*, **244**, 40 (1981).
188. Lubimov, V. A., et al., *Phys. Letts.*, **94B**, 266 (1980).
189. Reines, F., et al., *Phys. Rev. Letts.*, **45**, 1307 (1980); Barger, V., et al., *Phys. Letts.*, **93B**, 194 (1980).
190. Gunn, J. E., et al., *Astrophys. J.*, **223**, 1015 (1978); Zel'dovich, Ya. B., et al., *Pis'ma Astron. Zh.*, **451**, 457 (1980); Schram, R., Steigmon, G., *Astrophys. J.*, 243, 1 (1981); Bisnovaty, G. S.-Kogan and Novikov, I. D., *Astron. Zh.*, **57**, 899 (1980).
191. Reasenberg, et al., *Astrophys. J.*, (*Letts.*), **234**, L219 (1979); Will, C. M., *Proc. Roy. Soc. A.*, **368**, 5 (1979).
192. (a)Braginsky, V. B., *Usp. Fiz. Nauk*, **132**, 387 (1980).
192. (b) Berstein, I. M., and Shvartsman, V. F., *Zh. Eksp. Teor. Fiz.*, **79**, 1617 (1980).
193. Bekenstein, J. D., and Meisels, A., *Astrophys. J.*, **237**, 342 (1980).
194. Turner, M. S., and Schramm, D. N., *Phys. Today*. **32**, No. 9, 42 (1980); Dolgov, A. D. and Zel'dovich, Ya.B., *Rev. Mod. Phys.* **53**, 1 (1981).
195. Linde, A. D., *Phys. Letts.*, **92B** 119 (1980); **92B** 327, 394 (1980); Gurovich, V. Ts. and Starobinsky, A. A., *Zh. Eksp. Teor. Fiz.*, **77**, 1683 (1979); see also *Phys. Letts.*, **91B**, 99 (1980).
196. Kirzhnits, D. A., and Linde, A. D., *Priroda*, No. 11, 20 (1979).
197. Kighuchi, M., *Progr, Theor. Phys.* **63**, 146 (1980).
198. Nanopoolos, D. V., and Weinberg, S., *Phys. Rev.*, **D20**, 2484 (1979); Barrow, J. D., *Mon. Not. RAS*. **192**, 19 (1980).
199. Rees, M. J., *Contemp. Phys.*, **21**, 99 (1980).
200. Gurzadyan, V. G., and Ozernoy, L. M., *Pis'ma Astron. Zh.*, **5**, 630 (1979); Duncan, M. J. and Wheeler, J. C., *Astrophys. J.* (*Letts.*), **237**, L 27 (1980).
201. *Origin of Cosmic Rays*. IUPAP/IAU Symposium No. 94, Bologna, Italy, D. Reidel Publishing Co., 1981.
202. Mezets, E. P., et al., *Nature*, **282**, 587 (1979); Terrel, J., et al., *Nature*, **285**, 383 (1980).
203. Ramaty, R., et al., *Nature*, **287** 122 (1980).
204. Mazets, E. P., et al., *Pis'ma Astron. Zh.* **6**, 706 (1980); *Nature*, **290**, 378 (1981).
205. *Nature*, **284**, 507 (1980).
206 Bahcall, J. M., *Space Science Rev.*, **24**, 227 (1979).
207. Berezinsky, V. S., and Ginzburg, V. L., *Mon. Not. RAS*, **194**, 3 (1981); Berezinsky, V. S., *Usp. Fiz. Nauk* **133**, 545 (1981)
208. Ginzburg, V. L., *Voprosy filosofii*, No. 12, 24 (1980).
209. Kuhn, T. S., *The Structure of Scientific Revolutions*. Second Edition. The University Chicago Press, Chicago, (1970).
210. Fitch, V. L., *Rev. Mod. Phys.*, **53**, 367 (1981); Cronin, J. W., *Rev. Mod. Phys.*, **53**, 373 (1881).
211. Carlsson, A. E. and Ashcroft, N. W. *Phys., Rev. Lett.* **50**, 1305 (1983).
212. *Physics Today*, **36**, No. 4, 17 (1983,.
213. Goldbole, R. M., Pakvasa, S. and Roy, D. P., *Phys. Rev. Lett.*, **50**, 1539 (1983).
214. Goldhaber, M., *Physics Today*, **36**, No. 4, 35 (1983).
215. Schramm, D. N. *Physics*, *Today*, **36**, No. 4, 27 (1983).
216. *Physics Today*, **36**, No. 5, 17 (1983); Hawking, S., Gibbons, G. W., and Siklós S., (Eds.), *The Very Early Universe*, Cambridge University Press (1983).
217. UA1 Collaboration, *Phys. Lett.* **126B** 398 (1983).
218. Ginzburg, V. L. and Ptuskin, V. S., *T. Astophyl. Astron. (India)* **5**, 99 (1984)

Index